PATENTLY MATHEMATICAL

Patently
MATHEMATICAL

Picking Partners,
Passwords,
and Careers
by the Numbers

JEFF SUZUKI

JOHNS HOPKINS UNIVERSITY PRESS
BALTIMORE

© 2019 Johns Hopkins University Press
All rights reserved. Published 2019
Printed in the United States of America on acid-free paper
2 4 6 8 9 7 5 3 1

Johns Hopkins University Press
2715 North Charles Street
Baltimore, Maryland 21218-4363
www.press.jhu.edu

Library of Congress Cataloging-in-Publication Data

Names: Suzuki, Jeff, author.
Title: Patently mathematical : picking partners, passwords,
and careers by the numbers / Jeff Suzuki.
Description: Baltimore : Johns Hopkins University Press, 2018. | Includes
bibliographical references and index.
Identifiers: LCCN 2018010713| ISBN 9781421427058 (hardcover : alk. paper) |
ISBN 9781421427065 (electronic) | ISBN 1421427052 (hardcover : alk. paper)
| ISBN 1421427060 (electronic)
Subjects: LCSH: Mathematics—Formulae. | Patents.
Classification: LCC QA41.S9395 2018 | DDC 510.2/72—dc23
LC record available at https://lccn.loc.gov/2018010713

A catalog record for this book is available from the British Library.

Special discounts are available for bulk purchases of this book. For more information,
please contact Special Sales at 410-516-6936 or specialsales@press.jhu.edu.

Johns Hopkins University Press uses environmentally friendly
book materials, including recycled text paper that is composed of at least
30 percent post-consumer waste, whenever possible.

TO ED W. AND JEFF K.,
who listened patiently while I pontificated about mathematics

CONTENTS

ACKNOWLEDGMENTS

I'd like to thank Tom Clare, a US patent attorney at Baker and Hostetler, LLC, for guidance through the patent system. Tom's patient explanations of the fine points of patentability have proven invaluable; any mistakes or misinterpretations of patent law are my own. This book is intended for informational purposes only and not for the purpose of providing legal advice; those seeking legal advice should consult a professional attorney.

PATENTLY MATHEMATICAL

My Billion-Dollar Blunder

In my opinion, all previous advances in the various lines of invention will
appear totally insignificant when compared with those which
the present century will witness. I almost wish that I might live my life
over again to see the wonders which are at the threshold.

CHARLES HOLLAND DUELL (1850–1920), US COMMISSIONER OF PATENTS
Interview August 9, 1902

I could have invented Google.

Lest the reader think that this is a statement of monstrous vanity, let me qualify it. Google's search engine is based on mathematics principles that are among the first taught to math majors. *Any* mathematician could have invented Google.

This observation underscores a crucial fact about the modern world. The motto of every inventor is "Build a better mousetrap, and the world will beat a path to your door." But people have been improving mousetraps for thousands of years, so all of the obvious solutions have been found. The garage workshop, with a lone genius struggling to create a world-changing invention, is mostly a thing of the past. Applying the laws of science to develop a more effective painkiller, a faster microchip, or a more efficient solar cell takes the resources of an industrial giant and a laboratory where hundreds or even thousands of scientists and technicians work together.

Without some hope of ultimate compensation, few would make the effort. The patent system exists to promote innovation by giving inventors an opportunity to recover their expenses and to profit from their work. This is reasonable for patents based on physical discoveries. The production of electricity from sunlight relies on a fine balance of a large number of factors, and simultaneous independent discovery of the same mechanism is extremely unlikely. If a company finds a competitor is producing a solar cell identical to one of its own products, it is more reasonable to suspect industrial espionage or reverse engineering than it is to assume their research division found exactly the same solution.

But what of inventions based on mathematics? Applying the laws of mathematics requires nothing more sophisticated than a pencil and nothing more expensive than a paper notebook. History shows that simultaneous discovery of an idea is common in mathematics, and while a physical laboratory can provide all manner of paperwork to establish priority of invention, a mathematician's laboratory is in their head and rarely well documented. Thus, it is a long-standing principle of patent law that mathematical discoveries cannot be patented.

Yet, in recent years, patents based on mathematics have been issued by the thousands, based on a guideline that, if the mathematical process is connected to a device, the whole is patentable. You can't patent a mathematical algorithm, but you can patent a machine that runs the algorithm.

As a mathematician, I look on these patents with mixed emotions. After all, I *could* have invented Google. At the same time, mathematics is so broad a field that surely there are patentable—and profitable!—inventions just waiting to be made. However, allowing patents based on mathematics raises new problems.

Patent policy is directed toward promoting innovation. Patents must be for new and non-obvious inventions and cover the invention, not its uses: this encourages inventors to create new things and not find new uses for old things. At the same time, they must be limited in scope; otherwise, a too-broad patent could potentially be used in attempts to extort licensing fees from other innovators, stifling their creativity.

But the heart and soul of mathematics is its generalizability: the finding of new uses for old things. Therefore, granting a patent for an invention based on mathematics runs the risk of stifling innovation, not promoting it. It follows that patents based on mathematics must be carefully constructed.

In the following pages, I will guide the reader through the mathematics behind some recent patents. We'll see how some very simple mathematical ideas form the basis for a very broad range of patentable inventions. We'll also see how these same ideas can be retasked to be used in different fields. In the end, we hope to offer some insight into the question, *Under what conditions should a device based on a mathematical algorithm be patentable?*

1

\\\\\\\\\\\\\\\\\\

The Informational Hokey Pokey

This World Encyclopedia . . . would be alive and growing and changing
continually under revision, extension and replacement from the
original thinkers in the world everywhere. Every university and research
institution should be feeding it. Every fresh mind should be brought
into contact with its standing editorial organization. And on the other
hand its contents would be the standard source of material for the
instructional side of school and college work, for the verification of
facts and the testing of statements—everywhere in the world.

H. G. WELLS
Lecture delivered at the Royal Institution of Great Britain, November 20, 1936

Near the end of World War II, Vannevar Bush (1890–1974), the wartime
head of the Office of Research and Scientific Development, described
a "future device for individual use,"* which he called the *memex*. The
memex could store and present an unlimited amount of data, and it allowed the user to permanently link information, so that (to use Bush's
examples) someone interested in the history of why the Turkish bow
proved so superior to the English longbow during the Crusades would
be able to find and link relevant articles on the subject. Bush described
how the memex would change civilization: the lawyer would have at his

*Bush, "As We May Think," 121.

instant disposal all his own experiences, as well as all those of his colleagues; the patent attorney would have immediate access to all issued patents, and information relevant to all of them; the physician would have access to millions of case histories and so on.

The modern internet has largely fulfilled Bush's dream of a web of knowledge and experience: the lawyer, through services like Lexis Nexis, can research cases dating back to the 1800s; the patent attorney can consult several databases of patents and related technical reports; the physician has MEDLINE. And so the layman interested in why the Turkish bow proved superior to the English longbow can consult the Wikipedia article on these weapons, whereupon he would discover that the English did not actually adopt the longbow until after the Crusades: Bush, while a brilliant engineer, was no historian.

Bags of Words

Bush realized that *having* the information wasn't enough: "The summation of human experience is being expanded at a prodigious rate, and the means we use for threading through the consequent maze to the momentarily important item is the same as was used in the days of square-rigged ships."* While Bush's memex or the hyperlink structure of the web allows us to link related documents, how do we *find* a document on a subject?

One solution is to create a directory or an index covering all known documents or at least as many as it is practical to index. In January 1994, Jerry Yang and David Filo, two graduate students at Stanford University, published a web page named Jerry and David's Guide to the World Wide Web. This directory was sorted by content and contained links to other websites. The next year they incorporated as Yahoo!

Producing an index or directory is a challenging task. Consider a similar but far simpler problem: creating an index for this book. To do that, we need to create a list of index entries, which are the terms index users are likely to use. Each index entry then includes a reference to a page. Someone interested in looking something up finds a suitable index term and then goes to the referenced pages to find information about that term.

*Bush, "As We May Think," 112.

For example, this page seems relevant to someone looking for information about Yahoo!, so we'd make "Yahoo!" an index entry, and that index entry would include a reference to this page. Likewise, because Yahoo! was developed by Jerry Yang and David Filo, "Jerry Yang" and "David Filo" should also be index entries, as should "Stanford students" and "Jerry and David's Guide to the World Wide Web."

The problem indexers face is they must anticipate every possible use of the index by persons they have never met. For example, someone throwing a birthday party for someone born in January might want to know what happened in that month. Then, the fact that Yahoo! began in that month is relevant, so there should be an index entry "January" and a reference to this page. By a similar argument, index entries should also exist for "1994" and even for "index."

From this example, it should be clear that even a short piece of text will produce many index entries; the index entries for any given piece of text will generally exceed the length of the text itself. Because the index must anticipate the needs of everyone who will ever use the index, an indexer must be thorough.

The indexer needs to be consistent. While we introduced Jerry Yang and David Filo by giving their full names, how should we index them? The standard method would be to use index entries "Yang, Jerry" and "Filo, David." But someone unfamiliar with this convention might instead produce index entries "Jerry Yang" and "David Filo." And if Yang and Filo are mentioned later, it will almost certainly be by their last names only, so a human indexer would have to recognize that their names should not be indexed as a unit — "Yang and Filo"; instead, additional separate index entries for "Yang, Jerry" and for "Filo, David" are needed.

An indexer must be fast. While Yang and Filo were able to do the initial indexing of the web by hand, soon they had to employ an army of human indexers. But the web grew so rapidly that the human indexers fell farther and farther behind. Yahoo! stopped using human indexing in 2002, when there were about 40 million websites worldwide. Today, there are more than a billion websites and reportedly more than half a million are created every day. More than a trillion documents can be accessed through the internet.

Although humans are bad at thoroughness, consistency, and speed, computers excel at these things. Unfortunately, computers are bad at one crucial feature of indexing: in many cases, what something is *about* might not be evident from the text itself. Thus, in our discussion of Yahoo!, there was not one word or phrase (until this sentence) referencing that Yahoo! became one of the first *search engines*. A human indexer would recognize this, produce an index entry for "search engine," and include a reference to the preceding pages. But how can a computer be designed to do the informational hokey pokey, and find what a document's all about?

In the 1960s, a group led by Gerard Salton (1927–1995), a Cornell University professor, took the first steps in this direction. Salton's group developed SMART (System for the Mechanical Analysis and Retrieval of Text), which helped users locate articles on specific topics from a database from such diverse sources as journals on aeronautics and information theory and *Time* magazine.

Salton's team used *vectors*. Mathematically, a vector consists of an ordered sequence of values called *components*. Any systematic set of information about an object can be written as a vector. For example, when you visit a physician, several pieces of information are recorded. To record this information in vector form, we choose an order to list the components. We might decide that the first component is a patient's sex, the second component is the patient's height in inches, the third component is the patient's age in years, the fourth is the patient's weight in pounds, and the fifth and sixth components give the patient's blood pressure. Once we know this ordering, then we know vector [Male, 71, 38, 135, 115, 85] corresponds to a male patient who is 71 inches tall and 38 years old, who weighs 135 pounds and has a blood pressure of 115 over 85.*

Turning a document into a vector relies on the *bag of words* model. In this approach, we choose a set of *tokens*, usually keywords ("Chicago") or phrases ("the windy city"), and agree on some order. Roughly speaking, the tokens will be the keywords or phrases that we will be able to search for, so the tokens will be determined by how we intend to search our collection and limited only by our willingness and ability to deal with large vectors. For example, we might pick 10,000 commonly used words as our tokens and then convert every document into a 10,000-component *document vector* whose components correspond to the number of times a given token appears in the document.

*It's often convenient to have the vector components be numbers. We can do this by assigning numerical values to categorical data: thus, "Male = 1, Female = 2" or "Yes = 1, No = 0." Thus, our vector would be [1, 71, 38, 135, 115, 85].

For example, we might choose the tokens "I," "welding," "physics," "school," and "position." Then we could describe each document as a five-component vector, with each component giving the number of appearances of the corresponding keyword.

Now take three documents, say, the work histories of three people. Mercedes might write the following:

> *After I graduated from welding school, I took a job welding light- and medium-duty vehicles for a local company. My next position included welding shafts, cams, spindles, hubs, and bushings on autos, playground equipment, and trucks. I'm now an automotive welder at Ford.*

In contrast, Alice might have written this paragraph:

> *Graduated from Stockholm University with an undergraduate degree in physics, then went on to earn a PhD in high energy physics, then took a postdoc in physics at the Max Planck Institute for Physics in Munich. I then joined CERN as a Marie Curie fellow working on problems in high energy physics, then later took a senior position with the center's particle physics staff studying the physics of condensed matter.*

Finally, Bob might write the following:

> *Dropped out of welding school; got an undergraduate degree in physics; went to graduate school, also for physics; taught high school physics for several years before obtaining a position as a physics postdoc; currently seeking research job in physics.*

In Mercedes's paragraph, the word "I" appears 3 times; the word "welding" appears 3 times; the word "physics" appears 0 times; the word "school" appears 1 time; and the word "position" appears 1 time, so the document vector for her paragraph would be [3, 3, 0, 1, 1]. Meanwhile, in Alice's paragraph, the word "I" appears 1 time; the word "welding" appears 0 times; the word "physics" appears 7 times; the word "school" appears 0 times; the word "position" appears 1 time, so the document vector for her paragraph would be [1, 0, 7, 0, 1]. Finally, in Bob's paragraph, the word "I" appears 0 times; the word "welding" appears 1 time; the word "physics" appears 5 times; the word "school" appears 3 times; the word "position" appears 1 time, so the document vector for his paragraph would be [0, 1, 5, 3, 1].

Cosine Similarity

Suppose we construct the document vectors for everything in our library. Our next step would be to determine which documents are similar to one another.

A common way to compare two vectors is to find the *dot product* (also known as the *scalar product*, or the *inner product*): the sum of the products of the corresponding components of the two vectors. If two documents have no words in common, then the dot product will be 0 (because no matter how many times a keyword appears in one document, this will be multiplied by 0, the number of times the keyword appears in the other). In general, the higher the dot product, the more similar the two documents.

For example, if we take the document vectors for the work history paragraphs of Mercedes and Alice, which are $[3, 3, 0, 1, 1]$ and $[1, 0, 7, 0, 1]$, respectively, the dot product would be

$$3 \times 1 + 3 \times 0 + 0 \times 7 + 1 \times 0 + 1 \times 1 = 4$$

Meanwhile, the dot product of Alice's document vector work history paragraph with Bob's, which was $[0, 1, 5, 3, 1]$, will be

$$1 \times 0 + 0 \times 1 + 7 \times 5 + 0 \times 3 + 1 \times 1 = 36$$

This suggests that Alice's work history has more in common with Bob's than with Mercedes's.

There's one complication. Suppose we were to take Mercedes's paragraph and put 10 copies of it into a single document. Because of the document vector counts the number of times a given keyword appears, then every keyword will be counted 10 times as often. The new document will correspond to the vector $[30, 30, 0, 10, 10]$, and the dot product between Mercedes's revised paragraph and Alice's original paragraph is now

$$30 \times 1 + 30 \times 0 + 0 \times 7 + 10 \times 0 + 10 \times 1 = 40$$

Now it appears that Alice's work history has more in common with Mercedes's than with Bob's. But, in fact, Mercedes's work history hasn't changed; it's just been repeated.

To avoid this problem, we can *normalize* the vectors. There are several ways to do this. First, because the document vector records the *number* of times a given keyword appears, we can normalize it by dividing by the total number of times the keywords appear in the document. In Mercedes's repeated work history, with vector [30, 30, 0, 10, 10], the keywords appear 30 + 30 + 0 + 10 + 10 = 80 times. The value 80 is called the *L1-norm*, and we can normalize the vector by dividing each component by the L1-norm. Thus, [30, 30, 0, 10, 10] becomes L1-normalized as [0.375, 0.375, 0, 0.125, 0.125]; we call this a *frequency vector*.

There are two advantages to using the frequency vector. First, it's easy to compute: we need to find the total number of tokens in the document and then divide each component of the vector by this total. Second, the frequency vector has a familiar interpretation: they are the fractions of the tokens corresponding to any individual keyword. Thus, 37.5% of the keyword words in Mercedes's work history document are "I," while 12.5% are "school."

However, what is easy and familiar is not always what is useful. If we want to use the dot product to gauge the similarity of two vectors, then the dot product of a normalized vector *with itself* should be higher than the dot product of the normalized vector with any other normalized vector.

Consider Bob's vector [0, 1, 5, 3, 1], with L1-norm 0 + 1 + 5 + 3 + 1 = 10. The L1-normalized vector is [0, 0.1, 0.5, 0.3, 0.1], and the dot product of this vector with itself will be

$$0 \times 0 + 0.1 \times 0.1 + 0.5 \times 0.5 + 0.3 \times 0.3 + 0.1 \times 0.1 = 0.36$$

Now consider Alice's vector [1, 0, 7, 0, 1], with L1-norm 1 + 0 + 7 + 0 + 1 = 9 and corresponding L1-normalized vector [0.11, 0, 0.78, 0, 0.11]. The dot product of this vector with the first will be

$$0 \times 0.11 + 0.1 \times 0 + 0.5 \times 0.78 + 0.3 \times 0 + 0.1 \times 0.11 = 0.4$$

The dot product is higher, which suggests that Bob's work history is more similar to Alice's than it is to Bob's own.

To avoid this paradox, we can use the *L2-norm*: the square root of the sum of the squares of the components. This originates from a geometric interpretation of a vector. If you view the vector as a pair of numbers giving a set of directions, so that a vector [3, 7] would be "Go 3 feet east and 7 feet north," then the square root of the sum of the squares of the components gives you the straight line distance between your starting and ending points. In this case, $\sqrt{3^2+7^2} \approx 7.616$, so you've gone 7.616 feet from your starting point. The normalized vector is found by dividing each component by the L2-norm: thus, the vector [3, 7] becomes L2-normalized as [0.3939, 0.9191]. As before, if we treat the vector as a set of directions, these directions would take us 1 foot *toward* the ending point. Likewise, the vector [1, 9] becomes L2-normalized as [0.1104, 0.9939].

In general, the dot product of a L2-normalized vector with itself will be 1, which will be higher than the dot product of the L2-normalized vector with any other L2-normalized vector. If we want the dot product to be a meaningful way of comparing two vectors, we'll need to use the L2-normalized vectors. In this case, the dot product will give us the cosine similarity of the two vectors.*

The geometric interpretation becomes somewhat strained when we move beyond three-component vectors; however, we can still compute the L2-norm using the formula. For Bob's work history vector, we'd find the L2-norm to be $\sqrt{0^2 + 1^2 + 5^2 + 3^2 + 1^2}$ = 6, giving us L2-normalized vector [0, 0.1667, 0.8333, 0.5, 0.1667]. Meanwhile Alice's work history vector would have L2-norm $\sqrt{1^2 + 0^2 + 7^2 + 0^2 + 1^2}$ ≈ 7.14, giving us L2-normalized vector [0.1400, 0, 0.9802, 0, 0.1400]. The dot product of Bob's normalized vector with itself will be 1 (this will always be true with the L2-normalized vectors); meanwhile, the dot product of Bob's L2-normalized vector with Alice's L2-normalized vector will be

$$0 \times 0.1400 + 0.1667 \times 0 + 0.8333 \times 0.9802 + 0.5 \times 0 + 0.1667 \times 0.1400$$
$$= 0.8402$$

This fits our intuition that Bob's work history should be more similar to Bob's work history than to Alice's work history.

Returning to the work histories of Mercedes, Alice, and Bob, we'd find that every term of Mercedes's repeated work history, with document vector [30, 30, 0, 10, 10], would be divided by

$$\sqrt{30^2 + 30^2 + 0^2 + 10^2 + 10^2} \approx 44.72,$$

giving normalized vector [0.6708, 0.6708, 0, 0.2236, 0.2236].

Now, if we compute the dot products of the normalized vectors, we get (between Mercedes and Alice)

*Those familiar with trigonometry might recognize the term *cosine* as a mathematical function of an angle in a triangle. Like the L2-norm itself, the association of the dot product with the cosine originates from the geometric interpretation of a vector as a set of directions.

$$0.6708 \times 0.1400 + 0.6708 \times 0 + 0 \times 0.9802 + 0.2236 \times 0$$
$$+ 0.2236 \times 0.1400 \approx 0.1252$$

which confirms our sense that Alice's work history is more similar to Bob's than it is to Mercedes's.

It's tempting to proceed as follows. First, find the L2-normalized document vector for every item in our collection. Next, compute the dot product between each normalized vector and every other normalized vector; if the cosine similarity exceeds a certain value, identify the two documents as being on the same topic.

Could we have obtained a patent for this method of comparing documents? Strictly speaking, a patent cannot be granted for a mathematical idea, such as the dot product. Nor can you receive a patent for an application of the dot product, like using it to compare two documents to determine whether they are similar. However, a patent could be granted if that mathematical idea is incorporated into some machine. Thus, you could obtain a patent for a machine that compares documents using the dot product.

Unfortunately, there are two problems. First, the number of components of the normalized vectors will equal the number of distinct keywords, and it's not uncommon to use 100,000 or more keywords. Meanwhile, the number of document vectors will equal the number of documents. For indexing purposes, a document could be anything from a book to a single article in a newspaper magazine; with this in mind, it wouldn't be unreasonable for a physical library to have more than 10,000,000 documents. This means that we'll need to store and manipulate $100{,}000 \times 10{,}000{,}000 = 1{,}000{,}000{,}000{,}000$ numbers: roughly 1,000 gigabytes of data. Moreover, we'd need to find the dot product of *every* document vector with every other document vector: this requires finding $10{,}000{,}000 \times 10{,}000{,}000$ dot products, each requiring $100{,}000 \times 100{,}000$ multiplications: a staggering 1,000,000,000,000, 000,000,000,000 calculations. A computer that could perform a trillion computations each second would require more than 30,000 years to find all these dot products.

To be sure, computers are getting faster all the time. But there's a second problem: the dot product doesn't work to compare two documents. In particular, simple cosine similarity of document vectors does not adequately classify documents.

td-idf

Part of the problem is that not all keywords have the same discriminatory value. For example, suppose we make our keywords "the," "and," "of," "to," "bear," and "Paleolithic." One document vector might be [420, 268, 241, 141, 255, 0] and another might be [528, 354, 362, 174, 0, 213]. The dot product of the normalized vectors would be 0.88, suggesting the two documents are similar. However, if we look closer, we might discover that the first document mentions the word "bear" (the fifth keyword) hundreds of times and the word "Paleolithic" not at all, while the second document uses the word "Paleolithic" hundreds of times but omits the word "bear" entirely. Any sensible person looking at this information would conclude the two documents are very different.

Why does the cosine similarity fail? You might suspect that, because it ignores word order, cosine similarity would fail to detect the nuances of a text. However, the more fundamental problem is that words such as "the," "and," "of," and "to" are ubiquitous, and make up about 6%, 4%, 3%, and 1% of the words in a typical English document, so in some sense, most of the similarity between the two documents comes from the fact that both are in English. Remember our goal is to try and classify documents based on their word content. A word that appears frequently in a document may give us insight into the content of the document, but if the word appears frequently in *all* documents, it's less helpful: if you want to find a particular car, it's more useful to know its color than the number of tires it has.

In the early days of information retrieval, ubiquitous words like "the," "and," "of," "to," and so on were compiled into *stop lists*. A word on the stop list would be ignored when compiling the document vectors. However, creating a stop list is problematic: When do we decide a word is sufficiently common to include it? And when does a word become sufficiently *uncommon* to make it worth indexing? The difficulty is that common and uncommon depend on the documents we're indexing: *hyponatremia* is an uncommon word in a collection of Victorian literature but not in a journal on vascular medicine.

Moreover, excluding stop words from consideration limits the uses of the document vectors. Thus, while stop words are useless for indexing, several studies suggest that they provide a literary "fingerprint" that helps identify authorship. And because the internet links the entire world together, the prevalence of the word "the" on a web page is a clue that the web page is in English and not some other language.

Thus, instead of ignoring common words, we can limit their impact by replacing the frequency with the *term frequency-inverse document frequency*. This is the product of two numbers: the *term frequency* (tf) and the *inverse document frequency* (idf).

There are several ways of calculating the term frequency, but one of the more common is to use the *augmented frequency*. If a keyword appears k times in a document, and no keyword appears more than M times in the document, then the augmented frequency of that keyword will be $\frac{1}{2} + \frac{k}{2M}$. In the first document, the keyword "the" appears 420 times, more than any other keyword; thus, its augmented frequency will be $\frac{1}{2} + \frac{420}{2 \times 420} = 1$. Meanwhile, the word "bear" appears 255 times, so its augmented frequency will be $\frac{1}{2} + \frac{255}{2 \times 420} = 0.804$. Note that a word that doesn't appear at all would have an augmented frequency of $\frac{1}{2}$ (hence, "augmented").

What about the inverse document frequency? Suppose we have a library of N documents, and a particular word appears in k of the documents. The most common way to find the value of the inverse document frequency is to use $\log(N/k)$, where log is a mathematical function. For our purposes, it will be convenient to define the log of a number as follows: Suppose we begin with 1, and double repeatedly:

$$1 \to 2 \to 4 \to 8 \to 16 \to 32 \to \ldots$$

The log of a number is equal to the number of times we must double to get the number. Thus, if a word appears in 5 out of 40 documents, its idf will be $\log(40/5) = \log(8)$, and because 8 is the third number in our sequence of doubles, $\log(8) = 3$. Note that since we're basing our logs on the number of doublings, we say that these are *logs to base 2*.[*]

What about $\log(10)$? Since 3 doublings will take us to 8, and 4 doublings will take us to 16, then $\log(10)$ is between 3 and 4. With calculus, we can generalize the concept of doubling; this allows us to determine $\log(10) \approx 3.322$.

In the previous example, the word "the" appears in 2 out of 2 documents, so its idf would be $\log(2/2) = \log(1) = 0$ (because we need to

---

[*]Strictly speaking, we should write $\log_2(8) = 3$ to indicate our base is 2. However, as long as we agree to use no base other than 2, then we can omit the subscript.

double 1 zero times to get 1). Meanwhile, "bear" appears in 1 of the 2 documents, so their idf would be log(2/1) = log(2) = 1.

Now we can produce a vector using the tf-idf vector for our keywords. Consider our first document, with document vector [420, 268, 241, 141, 255, 0]. The largest component is the first, at 420: no keyword appeared more than 420 times. The tf for the first keyword will then be $\frac{1}{2} + \frac{420}{2 \times 420}$ = 1. Meanwhile, its idf, since it appeared in both documents, will be log(2/2) = 0. Thus, the first component of the tf-idf vector will be 1 × 0 = 0. The tf for the second keyword, which appeared 268 times, will be $\frac{1}{2} + \frac{268}{2 \times 420}$ ≈ 0.82, but it also appeared in both documents, so its idf will also be 0. The second component of the tf-idf vector will be 0.82 × 0 = 0. Continuing in this fashion, we obtain the tf-idf vector for the first document: [0, 0, 0, 0, 0.80, 0]. If we perform a similar set of computations for the second document, we find the weighted document vector will be [0, 0, 0, 0, 0, 0.70]. Normalizing and taking the dot product, we find the cosine similarity of the two documents is 0: they are nothing alike. Using tf-idf helps to limit the impact of common words and makes the comparison of the documents more meaningful. This is the basis for US Patent 5,943,670, "System and Method for Categorizing Objects in Combined Categories," invented by John Martin Prager of IBM and granted on August 24, 1999.

Latent Semantic Indexing

However, there's a problem. Suppose we have three documents in our library: first, an article on bears (the animal) that uses the common name "bear" throughout. The second article is identical to the first, but every occurrence of the word "bear" has been changed to "ursidae," the scientific name. The third article is about the Chicago Bears (a sports team). Here we see two problems with natural languages:

1. The problem of *synonymy*, where the same concept is expressed using two different words.
2. The problem of *polysemy*, where the same word is used to express two different concepts.

A giant leap forward in tackling the problems of synonym and polysemy occurred in the late 1980s when Scott Deerwester of the University of Chicago, Susan Dumais of Bell Communications, and Richard

Harshman of the University of Western Ontario in Canada developed *latent semantic indexing* (LSI; but also known as *latent semantic analysis, or LSA*). Together with Bell scientists George Furnas, Richard Harshman, Thomas K. Landauer (all psychologists), and Karen E. Lochbaum and Lynn A. Streeter (both computer scientists), they received US Patent 4,839,853 for "Computer Information Retrieval Using Latent Semantic Structure" on June 13, 1989. Again, the patent is not on the process of LSI/LSA, which is a purely mathematical construct, but on a device that applies LSI/LSA to a set of documents.

Consider the vectors for the documents in our library. We can compile all these document vectors into a *term document matrix*. We can view this as a table whose entries indicate the number of times a given keyword appears in a given document, where our rows correspond to the keywords and the columns to the documents. Thus, the value in the 15th row of the 385th column can be read as the number of times the 15th keyword appears in the 385th document.

If we simply look at the table, we see a jumble of information, and the fine structure is not apparent. However, we can apply a technique called *singular value decomposition* (SVD) to the term document matrix. The details of SVD are dauntingly technical, but the following analogy might be helpful. Imagine that you're at an amusement park. If you stand at the entrance and look into the crowd, the location of the people appears to be a random jumble. But if you look down on the crowd from above, you'll see that there are distinct clusters: around food stands, popular rides, and so on.

In effect, SVD finds a vantage point from which the clusters can be seen most easily. Documents within the same cluster are, in some sense, similar, so if we're able to identify the topic of *any* document in the cluster, then we can, with some assurance, conclude that *all* documents in the cluster are on the same topic. Therefore, a document about "bears" (as animals) would be in the same cluster as a document about "ursidae," but in a completely separate cluster from a document about the Chicago football team.

It's not clear how well LSI works. In 1993 and 1995, Dumais ran experiments on LSI and found that it routinely returned good lists of similar documents, suggesting it is a very promising way to catalog documents. But in 2005, Avinash Atreya and Charles Alkan of the University of California, San Diego, applied LSI to the documents in the TREC (Text Retrieval Conference) database and found that other methods of comparing documents, notably BM25 (based on tf-idf), performed better.

Regardless of whether LSI works, there's another problem: computing the SVD for even a modest-sized matrix is computationally intensive. In 1995, Dumais noted that finding the SVD for a collection of 80,000 documents with 60,000 keywords took between 18 and 20 hours. While modern computers are about 100 times faster than they were in 1995, the web has 100,000 times as many documents. Even worse, these documents are constantly changing. SVD is good for static collections of documents, but modern search engines need a faster way of indexing documents.

Nearest Neighbors

Roughly speaking, LSI circumvents the problem of synonymy and polysemy by focusing on the structure of the document itself: the words are less important. While effective, it takes an enormous amount of computer power to implement effectively, so we might try to tackle synonymy and polysemy directly.

One way is to use a thesaurus. For example, a thesaurus would indicate "bear" and "ursidae" are interchangeable terms. The best possible thesaurus would be produced by human beings, who know how other human beings use words. However, compiling a good thesaurus takes years, while spoken languages are living things, and words change their meanings over time. Tsar Ivan the Terrible of Russia (1530–1584) didn't get his nickname for being a bad ruler but because he inspired terror in the enemies of the state.

The problem of linguistic change is even more challenging for indexers, because several meanings might coexist, and they must anticipate which meanings a user will want — and that meaning changes with passing fancies. Today, a person might want information about "bears," expecting to be directed to articles about animals; tomorrow, that same person might expect information about the Superbowl and Chicago; the day after, the searcher would want articles on China and the stock market. It would be impossible for human indexers to keep up with the changes; what's necessary is a rapid and automated way to generate a thesaurus.

One such method is described in US Patent 5,675,819, "Document Information Retrieval Using Global Word Co-occurrence Patterns," invented by Hinrich Schuetze and assigned to Xerox Corporation on October 7, 1997. This patent illustrates one of the challenges confronting mathematicians and patent applicants: generalization is the heart and soul of mathematics, but patents must be for specific inventions.

An ordinary person might distinguish between a multivolume work on the natural history of bears, an article on bears in *National Geographic*, and a bumper sticker "I Brake for Bears." But to a mathematician, all of these are documents, so LSI could be applied to all of them. The difference is important for an inventor: because of the intentionally limited scope of a patent, it's possible to create a patentable device that uses the *same* mathematical process, as long as the machine it's connected to is sufficiently different. Schuetze's insight is that the bag of words model of documents can be applied to the keywords themselves.

Suppose our document consisted of just the keyword. The document vector will be mostly zeros, corresponding to the occurrences of the other keywords, with a single 1, corresponding to the keyword itself. For example, suppose we had just three keywords: "bear," "white," and "ursidae." The keyword "bear" would have document vector $[1, 0, 0]$; the keyword "white" would have document vector $[0, 1, 0]$; the keyword "ursidae" would have document vector $[0, 0, 1]$. We'll call these our *keyword vectors*.

Now consider a more traditional document, where the word "bear" appeared 35 times, the word "white" appeared 12 times, and the word "ursidae" appeared 1 time. The document vector would be $[35, 12, 1]$. However, it's convenient to think about this document vector as the *linear combination*

$$35\,[1, 0, 0] + 12\,[0, 1, 0] + 1\,[0, 0, 1]$$

where the *scalar multiplications*, $35\,[1, 0, 0]$, are performed by multiplying each component of the vector by the number 35 (producing the vector $[35, 0, 0]$), and the *vector addition* is performed by adding the corresponding components of the vectors, so that $[35, 0, 0] + [0, 12, 0] + [0, 0, 1] = [35 + 0 + 0, 0 + 12 + 0, 0 + 0 + 1] = [35, 12, 1]$. In effect, we're describing the document by noting it uses the keyword "bear" 35 times, the keyword "white" 12 times, and the keyword "ursidae" 1 time.

Why would we do this? Suppose we take the document and replace every occurrence of the word "bear" with "ursidae," the scientific name. Because the two words are synonyms, and we've changed nothing else about the document (we'll assume, for illustrative purposes, that our document only uses the word "bear" to indicate the animal), then we ought to view the two documents as being essentially the same document. However, the frequency vector would change to $[0, 12, 36]$, and the document would appear radically different.

To prevent this misleading occurrence, suppose we alter our keyword vectors. Instead of $[1, 0, 0]$, $[0, 1, 0]$, $[0, 0, 1]$, we use the *thesaurus vectors* $[1, 0, 1]$, $[0, 1, 0]$, and $[1, 0, 1]$. Now consider the original document, where the word "bear" appeared 35 times, the word "white" appeared 12 times, and the word "ursidae" appeared 1 time. It would be represented using the linear combination

$$35\,[1, 0, 1] + 12\,[0, 1, 0] + 1\,[1, 0, 1] = [36, 12, 36]$$

Meanwhile, the revised document, where "bear" occurs 0 times, "white" appears 12 times, and "ursidae" appears 36 times, would be represented using the linear combination

$$0\,[1, 0, 1] + 12\,[0, 1, 0] + 36\,[1, 0, 1] = [36, 12, 36]$$

Even though the words of the document have changed, the document vector has not, reflecting the fact that the change of words has not affected the content of the document.

Why did this work? Note that we've assigned "bear" and "ursidae" the same keyword vector: $[1, 0, 1]$. This means that, whether the word appears as "bear" or as "ursidae," it contributes the same amount to the final document vector.

It should be clear that we can assign identical thesaurus vectors to exact synonyms and that keywords with similar meanings ought to be assigned similar vectors: this helps resolve the problem of synonymy. But how do we assign such vectors in the first place? Schuetze, in the Xerox patent, outlines the procedure. Consider a word like "axolotl." If you don't know what this word means, you can look it up in a dictionary. But our neighbors help define us: if you see the word in context, you can infer a great deal about what it means. If we read "The axolotl is found in Lake Xochimilco, and eats worms, insects, and small fish," we might conclude the axolotl is some sort of water-dwelling animal.

This suggests the following approach. Suppose we take our keyword and form *nearest neighbor documents*, consisting of the keyword and its nearest neighbors. The patent itself suggests using the nearest 50 words (the 50 words before and the 50 words after the keyword), but for illustrative purposes, we'll look at smaller neighborhoods, say the 5 nearest words. Our nearest neighbor documents for the keyword "bear" might look like the following:

- "The natural diet of a bear includes berries, insects, honey, and"
- "But contrary to popular belief, bears do not hibernate in winter."
- "last night's game, the Chicago Bears beat the Kansas City Chiefs"
- "in Chinese financial markets a bear market ahead, unless the Fed"

and so on. We put these together to form a single document:

The natural diet of a bear includes berries, insects, honey, and But contrary to popular belief, bears do not hibernate in winter. last night's game, the Chicago Bears beat the Kansas City Chiefs in Chinese financial markets a bear market ahead, unless the Fed . . .

While this reads like a cut-and-paste term paper written by a lazy student, the bag of words model views it as a document, so we can construct its document vector. We can do the same thing for all our other keywords, weight them using a tf-idf scheme, and then normalize them.

We might simply use the document vector for our keyword *as* the thesaurus vector for that keyword. However, the Xerox patent suggests instead applying SVD to the term document matrix to find the thesaurus vectors. This might seem a paradox: after all, we're considering the keyword vectors as an alternative to performing SVD, because SVD is too computationally intensive. However, the problem isn't SVD, it's SVD on a large collection of documents. In this case, while our library might contain millions of documents, we're reducing these documents to 100,000 or so keyword documents. This can be done quickly (or at least, quickly enough so that the results aren't obsolete by the time the SVD is completed). Using SVD, we can reduce our 100,000 keyword vectors to a few thousand *thesaurus vectors*. We can then produce a *context vector* for any document by writing it as a linear combination of the thesaurus vectors, weighted using some tf-idf scheme.

There are several useful features of this approach. First, it takes care of synonymy, because any time a keyword is used in a document, the frequency of *all* keywords associated with it is increased. Thus, it doesn't matter if the document uses "bear" or "ursidae," as both would increase the word count of "bear" and "ursidae."

This approach also alleviates the problem of polysemy. Consider a document that contains the word "bear." Using our keyword vector, the frequency of several associated words would be raised, such as "forest," "football," and "market." However, if the document is about the animal, the frequency of "forest" would likely also be increased, both through the inclusion of the keyword forest itself, as well as the inclusion of other keywords where "forest" is in the keyword neighbor document.

This approach helps solve part of the memex problem: we can automatically analyze every document and identify what the document is about. But this raises a new problem. Any given topic may have thousands of documents relevant to it. Which ones are the most important? We'll take a look at this question in the next chapter.

2

〰〰〰〰〰〰〰

The Trillion-Dollar Equation

*The tank is a big buildin' full of all the facts in creation an' all the
recorded telecasts that ever was made — an' it's hooked in
with all the other tanks all over the country — an' everything you
wanna know or see or hear, you punch for it an' you get it.*

MURRAY LEINSTER (1896–1975)
"A Logic Named Joe"

Science fiction authors are often credited with predicting marvelous
devices such as spaceships and computers. But the devices themselves
are little more than technology-clad versions of fairy-tale wonders,
with seven-league boots, magic swords, and genies replaced by spaceships, light sabers, and robots. Fantasy can limit the wielding of power
to a select few, chosen arbitrarily and capriciously, while the powers
of science and technology can be used by anyone: once unleashed, the
genie can never be returned to the bottle, and society must change.

As an example, consider Murray Leinster's "A Logic Named Joe." The
short story, published in 1946, centered around a small household device — the Logic — that gives users access to all the world's knowledge
and allows them to find weather reports, sports scores, historical trivia,
and stock quotes, much like Bush's memex. But if all the world's data are
available, then in principle, one could find the answer to any question:
it's just a matter of putting the pieces together. Because of some random
factor in its construction, the eponymous Joe could do exactly that.

Leinster realized that if you could look up who won today's race at Hialeah, you could also look up details on how to commit the perfect crime, find personal information about your neighbors, stalk former lovers, and view risqué material without censorship. The narrator—we'd call him a computer technician—wryly observed "things was goin' to pot because there was too many answers being given to too many questions." The narrator saved the world from itself by disconnecting Joe from the network.

We might compare Leinster's satire with Bush's article. Bush's memex shows us the potential of the information age, while Leinster's Logic shows us the risks. But both predictions rely on a key technological advance: the ability to find relevant information among a mountain of data.

Opening Salvos in the Search Engine Wars

In the early days of the web, Yahoo! and similar sites provided directories so users could locate information online. However, directories require that the searcher and the indexer have similar ideas about how to organize information.

In 1997, Linda Katz of Garden City, Kansas, was putting together her personal web page. On a whim, she added a note: "Tumbleweeds for sale." It was intended as a joke: tumbleweeds are a ubiquitous feature of the plains states, and while someone might pay to have tumbleweeds removed, no one would pay to have them delivered.

A human indexer, coming across the site, might conclude that the notice was a joke and choose not to index it. But if they chose to index it, how would it appear as an index subject: under "tumbleweed"? "Plants for sale"? "Decorations"? While they could list it under several topics, they'd still have to make a sequence of decisions about topics and subtopics: for example, Retail > Plants > Dried > Tumbleweeds.

At the other end, consider a party planner in Japan trying to throw a party with an American Old West theme. They might try to find vendors, but their ability to find vendors using a directory requires they use the same terms as the directory makers. At best, they'd have to navigate through a complex scheme of topics and subtopics: Entertainment > Party > Party Supplies > Themes > United States > Old West.

It would only be through a remarkable coincidence that a party planner in Japan would find Katz's tumbleweeds. Yet it happened, and soon Katz began receiving orders for tumbleweeds from places as diverse and exotic as Japan, Austria, Sweden, and Hollywood. With nearly eight billion people on the planet, there's almost certain to be a sizable market for any product.*

The problem is connecting those who demand a good or a service with those who supply it. What's needed is a *search engine*. Modern search engines consist of three parts: a *web crawler*, which seeks out accessible web pages; a *back end*, which identifies the contents of the web page; and a *front end*, which connects a searcher's query with an appropriate document.

Intuitively, a web crawler is easy to design: Imagine a randomizer that sends the crawler to a random web page. If the web page has outgoing links, the crawler chooses one of them at random and goes to a new web page. If this web page has outgoing links, the process is repeated. However, if the web page has no outgoing links (a *sink*), the randomizer take the crawler to a new random web page. With sufficiently many crawlers and sufficient time, a large portion of the web can be visited this way.

The back end is more complicated and generally uses a scheme for identifying document contents; we discussed some of these schemes in chapter 1. A key point to note about the back end is that all of the processing is done "offline" before a user submits a query. Because web pages change constantly, this processing must be done more or less continuously.

The front end is more complicated still. The Holy Grail of information retrieval is the ability to process *natural language queries*, wherein searchers could enter questions conversationally: "Find me someone to fill a welding position."

Boolean Searches

The bag of words model allows us to treat *any* collection of words as a document. A query "Find me someone to fill a welding position" is itself a document, and we can convert it into a vector and then compare it

*Or behavior. In 2003, Peter Morley-Souter wryly summarized this as Rule 34 of the internet: if it exists, there's pornography about it on the internet.

to the vectors of documents we've processed to find the most relevant document (in this case, perhaps someone's work biography).

Suppose we've produced the document vectors for a large number of web pages. Remember that, in our document vectors, each component corresponds to the frequency of a keyword in the document: the greater the value of the component, the more prominently the document features the keyword.

Consider the query: "Find me someone to fill a welding position." A simple approach is to use a *Boolean search algorithm*. Since the document vectors tell us how often a given keyword appears in the document, we can identify the keywords in the query (perhaps "welding") and then select those documents that include the keyword.

Boolean searches are characterized by speed and accuracy: they quickly return a set of hits, every one of which has the required characteristic. Moreover, nothing that lacks the required characteristic will be included. The problem with a Boolean search is that it will generally return far too many hits, while missing others. For example, a Boolean search for documents that included the word "welding" would return documents like:

> *Although I got my welding license right after school, it was several years before I could get a welding job, due to my poor performance on the welding licensing exam. After a job site incident caused by my negligence, my welding license was revoked.*

where the word "welding" appears four times. It's generally not possible to structure a Boolean source to avoid listing these documents. At the same time, the Boolean search would miss a document like

> *I've worked as an automatic welder for 5 years, then as an arc welder for 10 years; I have been a Master Welder for 5 years.*

because this document does not contain the word "welding."

A Boolean search can be modified to find the second type of document, usually through the invocation of arcane incantations such as "grep [Ww]eld*."* However, there are two other problems with a Boolean search.

---

*The difference between this and magic is that even a muggle can use a UNIX shell command.

First, Boolean searches require the searcher to use the same keywords that appear in the document. However, studies show that even when people look for common topics, they rarely use the same keywords. One person might search for "Who won today's race at Hialeah," and another might search for "win place and show results Hialeah racetrack," and neither will be useful if the search engine uses the keywords "quarter horse race results."

Second, the problem is even worse, since most people use search engines to find things they *don't* know: "What's the name of the thing in a fraction that's between the thing on top and the thing on bottom?" This is a phenomenon known in the computer industry as GIGO (garbage in, garbage out) or PEBKAC (problem exists between keyboard and chair): What people want and what they ask for are often very different.

However, you have to start somewhere. The Boolean search will return a list of documents that might answer the searcher's query. What's needed is some form of *relevance ranking*, so that the documents most likely to answer what the searcher wants (instead of what they ask for) are near the top of the list.

The Most Important Words Never Read

Traditional document retrieval strategies were based on traditional document collections: university libraries, corporate files, and government publications. Deliberate misinformation was rare, and nothing would be gained by diverting searchers from one document to an irrelevant one.

In contrast, anyone can put up a web page, the contents of which are limited only by what the hosting site would publish. Radical and irrational views on history, economics, race, and politics could be made available, and automated document classification systems could correctly identify the subject without evaluating the accuracy of the content.

Consider the home page of a hate group, filled with rabid claims about the inferiority of other races. Most people, when confronted with such a page, would either reject the contents as nonsense or accept the contents without hesitation. In both cases, the decision is made on the basis of belief: either a belief that no race is inherently superior to another or a belief that one race is.

If we wanted to base our evaluation on something other than preconceived notions, we would look at the supporting evidence. A principle of evidence is that bald assertions are meaningless. Thus, if the website proclaimed "98% of Masurians abuse narcotics!" but gave no sources, this statement should be disregarded.

If the website supplies a source, we must evaluate the reliability of the source. If the information comes from the US Department of Health and Human Services, we might accept the claim; if the information comes from a group known for its anti-immigrant stance, we might suspect the numbers.

This suggests a way to evaluate a document: check the sources. If there are none, or the sources are dubious, the document is suspect; however, if the sources are good, we might consider the document a more reliable source of information.

If we judge a document by its sources, then how do we judge the sources? In the scientific and scholarly community, judgment of sources is done by consensus: trusted sources will be used more often than questionable ones. Almost anyone writing about information retrieval will refer to Salton's 1970 book, *A Theory of Indexing*. From this fact alone, we might conclude that, if we want to know about indexing, Salton's book is a good place to start and should have a high relevance ranking. Sources referred to by many others are called *authorities*. The counterpart to authorities are *hubs*: documents that refer to many sources. These correspond to books and articles with extensive bibliographies that direct researchers to other relevant documents.

On the web, we can obtain this type of information through *Hyperlink Induced Topics Search* (HITS). On March 7, 1997, Jon Michael Kleinberg, a computer scientist at Cornell University who was then a visiting researcher at IBM, submitted a patent application for "Method and System for Identifying Authoritative Information Resources in an Environment with Content-Based Links between Information Resources." This would be granted to IBM as US Patent 6,112,202 on August 29, 2000.

Figure 2.1 shows two simple sets of connected web pages. In Simplified Internet 1, we have three web pages A, B, and C, where A has a link to B and to C, B has a link to C, and C has a link to A. In Simplified Internet 2, we have three web pages, D, E, and F, where E and F link to D, but D links to nowhere.

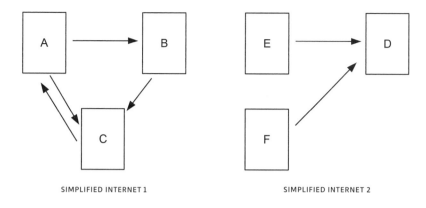

SIMPLIFIED INTERNET 1   SIMPLIFIED INTERNET 2

Figure 2.1. Sample Internet Neighborhoods

Connected systems like this are so important that several different disciplines have studied them. As a result, they're known by many names. Depending on which field you're working in, you might call it a web, a network, or a sociogram, though mathematicians usually use the term *graph* and include the study of such objects in the field of *graph theory*.

The graph consists of *nodes* (in this case, the web pages A, B, C), which are joined by *links* (the arrows). Here, the links are directional: you can go from A to B, but you can't go from B to A. Because of this, we say that we have a *directed graph*, or *digraph* for short. In other graphs, the links might be bidirectional: Our graph might show a group of people, where two people are linked if they are friends. In this case, if P is friends with Q, then Q is friends with P.

Imagine that our crawler is on page B, and our back end deems B relevant to a particular topic. The web crawler can follow outgoing links from B to other web pages. Some of these might be unrelated to the topic. However, a few web pages might be deemed related; for example, the back end might determine that web page C deals with a topic similar to that of web page B. The search engine can do the same thing at C and follow its outgoing links to find other relevant web pages. In this way, it can build up a network of web pages all on a given topic.

Now imagine a user queries the search engine for information about a topic. The search engine might simply present the user with a list of all pages on that topic. However, there may be hundreds, if not thousands, of associated web pages. In order for the search engine to return useful results, it would have to rank the pages in some fashion. One possibility is to determine an authority value for each page: the higher the authority value, the more trustworthy the page, and the higher the rank it receives.

Because a page's authority is determined by how many pages link to it, then it's also useful to compute the hub values of web pages as well. The authority values of the relevant pages can be put into an *authority vector*; in our Simplified Internet 1, the first, second, and third components of the authority vector correspond to the authority values of pages A, B, and C, respectively. In a similar fashion, the *hub vector* gives us the hub values of pages A, B, and C. Our goal is to find the authority vector.

To do this, we *initialize* the vectors by assigning them a starting value. One potential problem is that our initial values may alter our conclusions over which pages are the best hubs and which are the best authorities. To circumvent this difficulty, Kleinberg set the initial hub values of *all* pages to 1. This makes our initial hub vector $[1, 1, 1]$. Since we're not going to be comparing this vector to any other vector, it's not necessary to normalize it, though (as we shall see) it's convenient to scale it: the patent suggests dividing each component by the sum of the squares of the components. In this case, we'd divide each component by $1^2 + 1^2 + 1^2 = 3$, and so our initial hub vector will be $[\frac{1}{3}, \frac{1}{3}, \frac{1}{3}]$.

Now we compute an authority vector, where each component is the sum of all hub weights linking to the corresponding page.

1. The authority weight for page A will be $\frac{1}{3}$: only page C (hub weight $\frac{1}{3}$) links to it.
2. The authority weight for page B will be $\frac{1}{3}$: only page A (hub weight $\frac{1}{3}$) links to it.
3. The authority weight for page C will be $\frac{2}{3}$: pages A and B (each with hub weight $\frac{1}{3}$) link to it.

The authority vector will be $[\frac{1}{3}, \frac{1}{3}, \frac{2}{3}]$. The sum of the squares of the components is $\frac{6}{9}$, and if we divide the components by this value, we have the normalized vector $[\frac{1}{2}, \frac{1}{2}, 1]$.

Next, we find a new hub vector. This time, the new hub weights will be the sum of the authority weights the hub page links to. Thus:

1. The new hub weight of A will be $\frac{3}{2}$: it links to pages B (authority weight $\frac{1}{2}$) and C (authority weight 1).

2. The new hub weight of B will be 1: it links to page C (authority weight 1).
3. The new hub weight of C will be $\frac{1}{2}$: it links to page A (authority weight $\frac{1}{2}$).

Our new hub vector will be $[\frac{3}{2}, 1, \frac{1}{2}]$, which we normalize to $[\frac{3}{7}, \frac{2}{7}, \frac{1}{7}]$.

Now we repeat (*iterate*) this process and use the new hub vector to compute a new authority vector, which after normalization will be $[\frac{1}{5}, \frac{3}{5}, 1]$, which can then be used to find a new hub vector, $[0.4444, 0.2778, 0.0556]$. This new hub vector can be used to find a new authority vector $[0.0769, 0.6154, 1]$. We can use this new authority vector to find a new hub vector $[0.4468, 0.2766, 0.0213]$, which can then be used to find a new authority vector $[0.0294, 0.6176, 1]$, and so on.

It might not seem obvious how to use this information because the hub and authority values keep changing. However, the situation is akin to trying to predict the winner of the next presidential election by constantly taking polls. At the beginning of the campaign season, the support numbers for each candidate change rapidly, but after a while, the numbers settle down. In the same way, the hub and authority values might change considerably during the first few transitions, but once the numbers settle down, we might be able to say something about the significance of each web page.

The next few hub vectors will be

$$[0.4468, 0.2766, 0.0213] \rightarrow [0.4472, 0.2764, 0.0081] \rightarrow$$
$$[0.4472, 0.2764, 0.0031] \rightarrow \ldots$$

The values appear to getting closer to $[0.4472, 0.2764, 0]$, so we say that the hub vector *converges* to this *limit*. This suggests the hub values of pages A, B, and C should be 0.4472, 0.2764, and 0, making A the best hub and C the worst.

In a similar fashion, the next few authority vectors are

$$[0.0294, 0.6176, 1] \rightarrow [0.0112, 0.6180, 1] \rightarrow [0.0043, 0.6180, 1] \rightarrow$$
$$[0.0016, 0.6180, 1] \rightarrow \ldots$$

which also appears to converge to a limit: $[0, 0.6180, 1]$ in this case. This suggests that C is the best authority, while A is the worst.

The First Giant

Kleinberg's work received patent number 6,112,202. On the same date, August 29, 2000, the very next patent, 6,112,203, would be assigned to the first giant of the search engine industry: AltaVista.

AltaVista was a new player in the search engine market, having been launched on December 15, 1997. Many others had solved one problem: finding a way to classify a document and return documents relevant to a query. However, few had solved a much more difficult problem: making money.

The problem is that searching through millions of documents is a computationally intensive task and requires a battery of expensive machines. But while everyone knew the internet was a gold mine, no one had quite worked out how to make it profitable. Search engines were particularly problematic. You couldn't charge for access, since users would simply switch to free search engines. You might try to charge vendors, offering to rank their websites higher, but once users caught on to the biased nature of the searches, you'd risk losing market share. You might try to run advertising, but too many banner ads and surfers would go to other sites.

AltaVista solved the financing problem by being a project of the Western Research Laboratory of the Digital Equipment Corporation (DEC). Processing was handled using DEC's fastest machines, the Alpha processor, and in many ways, AltaVista was meant to be an advertisement for the power and speed of DEC computers; the fact that it ran internet searches was incidental.

AltaVista's search engine used an approach developed by DEC scientists Krishna Bharat and Monika Henzinger. As in Kleinberg's approach, Bharat and Henzinger began with a set of connected web pages. In general, if we find *every* page linked to a given set of pages and *every* page that links to a given set of pages, we will have a set of thousands or even millions of interconnected pages: too many to be handled easily. Thus, it's necessary to *prune* the set of pages.

Pruning occurred in several steps. First, for each page in the collection, we'd calculate

$$\text{Score} = \text{In-degree} + 2 \times (\text{URL hits}) + \text{Out-degree}$$

where "In-degree" is the number of web pages in our collection with links to the page, "Out-degree" is the number of web pages in our collection the page links to, and "URL hits" is the number of times the query words show up in the page's URL (Uniform Resource Locator, the human readable "address" of the web page, shown in the navigation bar).

For example, suppose we do a search for "cars," and we obtain an initial set of connected web pages. Suppose "carscarcars.com/cars_for_ sale" has links from 5 others in our set and links to 20 others. Then it has In-degree 5, Out-degree 20, and URL hits 4, which gives it a preliminary score of 5 + 2 × 4 + 20 = 30. Meanwhile, if "cars_for_sale.com/bargains" has 8 incoming links from other web pages in our set and outgoing links to 25 others, it will have score 8 + 2 × 1 + 25 = 35. We can compute such scores for all the pages and then present the searcher with the pages that have the highest scores.

Note that this method tends to reward companies that include keywords in their domain names: web pages at the domain "carscars carscarscarscarscars.com" would begin with an advantage over web pages at the domain name "cars.com." So many search engines use the URL as part of their ranking strategy that the practice of *cybersquatting* has become common: individuals and companies would buy up domain names, anticipating reselling them at a later point.

To avoid giving too great a weight to a website by its URL, additional pruning steps are necessary. First, the top scoring web pages might be analyzed for content, and additional pages might be discarded. Then the URL's effect on the web page's ranking is completely eliminated by computing 4 × In-degree + Out-degree, giving another score. Assuming that both "carscarcars.com/cars_for_sale" and "cars_for_sale.com/ bargains" are still in our list, their new scores would be 4 × 2 + 20 = 28 and 4 × 8 + 25 = 57. Note that this score favors authorities (which are linked to by many other pages and, hence, have high in-degree) over hubs (which have high out-degree).

This gives us a manageably small set of websites. To produce the final ranking, Bharat and Henzinger describe the following modification of Kleinberg's algorithm. In Kleinberg's algorithm, a hub weight for a page is computed by summing the authority weights of all pages the first page connects to. However, we can boost a hub weight by having multiple connections to what is essentially the same site: thus a page on "carscarscars.com" might link to "carscarscars.com/car1," "carscarscars .com/car2," "carscarscars.com/car3," and so on. To correct for this, the authority weights of these pages would be divided by the number of other pages from the same site. Likewise, when computing the authority weights, the hub weights are divided by the number of other hubs from the same site.

Finally, the pages are sorted by their hub and authority values. This allows us to go from thousands or even millions of web pages that *might* be relevant to a query to a few hundred web pages ranked by relevance.

Crowdsourcing Searches

AltaVista might have become the dominant search engine in the world but for two events. First, DEC made its fortune and reputation selling large computer systems to scientific and technical laboratories, and while DEC introduced a line of personal computers in 1982, they lacked the expertise to sell to businesses and individuals. The Alpha processor was part of a last-ditch effort to save the company by selling high-end workstations. It failed, and in June 1998, just six months after AltaVista launched, DEC would be acquired by Compaq computers. Unfortunately, Compaq suffered from several bad decisions (including the decision to buy DEC) and would be acquired by Hewlett-Packard. Each new owner had a different idea on how to make AltaVista profitable, and, in time, a strategy might have emerged, but time was the one thing search engine developers did not have.

On January 8, 1998, less than a month after AltaVista launched, Lawrence Page, working towards a PhD in computer science at Stanford University, submitted the application for what would become US Patent 6,285,999, "Method for Node Ranking in a Linked Database." A few months later, Page and Sergey Brin, another PhD student at Stanford, presented a paper describing a new search engine that, because they hoped it would be able to process hundreds of millions of web pages, they named after a very large number: 10100 (a 1 followed by 100 zeros), named a googol, though they chose to spell it *google*. The patent would be granted on September 4, 2001.

Page's insight began with a common way to gauge the importance of scientific papers: *citation counting*, the number of times the paper is cited by other papers. We might expect that a paper cited 25 times is more important than a paper cited 1 time. However, ranking by citation counting is overly simplistic: a reference to a paper from an obscure article would count just as much as a reference to a paper from an important article. Thus, we want to weight the articles themselves.

Kleinberg's algorithm, as well as the modified algorithm developed by Bharat and Henzinger, took this into account by determining the authority score of each page a hub linked to. However, there's a practical problem with this approach. Given any web page, it's easy enough to find every page it links to, so finding the exact hub value is possible. At the same time, finding the exact authority value is impossible: we'd need a complete map of the internet to find *every* page that links to the given page.

Instead, Page's algorithm ranked pages based on a model of how surfers navigate through the web: today, we might say that it makes use of crowdsourcing to rank web pages. Consider the Simplified Internet 1 of Figure 2.1, and suppose we track how many people are viewing each web page at any given point in time. We can then use the number of people viewing the web page as a measure of the importance of the web page.

To do this, we construct a *discrete time model*. Discrete time models are essentially pictures of a phenomenon, taken at specific intervals: a traffic camera might record the number of cars in an intersection at 11:45:30 and then at 11:45:35.

To construct a discrete time model, we describe the state of the system at some point in time; we often begin with an *initial state*. Next, we describe a *transition function* that tells us how to determine the state of the system at the next point in time from its current state. It's also convenient to keep track of the time step t.

In Page's approach, the state of the system corresponds to the number of surfers on each page. The basic transition function can be described simply: When we ring a bell, every person on a web page follows an outgoing link to an adjacent web page. If there are two or more outgoing links, traffic is divided equally among the possible destinations. Note that surfers only move one web page: thus, a surfer on page C could move to page A but not directly to page B. If we allow web surfers to continue to move around the internet, then the number of surfers on each page will (hopefully) converge to a limit, and the limit can be used to rank the pages.

Suppose we start at $t = 0$ with 1,000 surfers on web page A, and 0 persons on pages B and C. We can represent this situation using a *rank vector*: [1,000, 0, 0], where the three components represent the number of surfers viewing pages A, B, and C, respectively; because this is also our starting point, we can refer to it as our *initial value*. Now we apply the transition function and find the number of people on each page at $t = 1$.

There are several ways to do this, but as a general rule in life, it's easier to figure out where you've come from than where you're going. Therefore, we might determine where the people on each web page have come from.

Consider page A. The only way for surfers to get to page A is to start on page C. Moreover, A is the *only* destination possible for the surfers on C. Thus, the number of people on page A at $t = 1$ will be equal to the number of people on page C at $t = 0$. Since there were 0 people on page C at $t = 0$, there will be 0 people on page A at $t = 1$.

What about page B? Again, the only way for surfers to get to page B is to start at page A. However, B is one of two possible destinations for surfers on page A. Only half of the surfers on page A at $t = 0$ will go to page B. Since there were 1,000 people on page A at $t = 0$, there will be 500 people on page B at $t = 1$.

Finally, the surfers on page C at $t = 1$ come from two sources: (1) half of those on page A at $t = 0$ and (2) all of those on page B at $t = 0$. There will be $500 + 0 = 500$ surfers on page C at $t = 1$. Putting these values together, we get our new rank vector: $[0, 500, 500]$.

What if we apply our transition function a second time? We'd find the following:

- The surfers on A at $t = 2$ will be all surfers on C at $t = 1$: 500.
- The surfers on B at $t = 2$ will be half the surfers on A at $t = 1$: 0.
- The surfers on C at $t = 2$ will be half the surfers on A at $t = 1$, and all the surfers on B at $t = 1$: $0 + 500 = 500$.

This gives us rank vector is now $[500, 0, 500]$.

We can continue to apply the transition function. This will produce a sequence of rank vectors:

$$[500, 0, 500] \rightarrow [500, 250, 250] \rightarrow [250, 250, 500] \rightarrow$$
$$[500, 125, 375] \rightarrow [375, 250, 375]$$

If we keep going, we'll need to allow for fractions of a person, and our next rank vector will be $[375, 187.5, 437.5]$. Continuing in this fashion, we can produce as many rank vectors as possible, simulating the surfing behavior of 1,000 people over the three-page internet.

As with the hub and authority vectors, our rank vectors will begin to converge. The 20th iterate will be [399.414, 200.195, 400.391]. After that, we'll have

$$[399.414, 200.195, 400.391] \rightarrow [400.391, 199.707, 399.902] \rightarrow$$
$$[399.902, 200.195, 399.902]$$

Our rank vectors appear to be converging to [400, 200, 400].

Now consider the vector [400, 200, 400], and suppose we *began* with this rank vector: in other words, we started with 400 people on page A, 200 on page B, and 400 on page C. When we ring the bell and everyone moves, we find:

- Those on A will be all of those on C: 400.
- Those on B will be half of those on A: 200.
- Those on C will be half of those on A, together with all of those on B: 200 + 200 = 400.

The new rank vector will be [400, 200, 400] — the same as our original rank vector.

It's important to understand that surfers *have* moved around, so the 400 people on A after we ring the bell are different from the 400 on A before we rang the bell. But because the comings and goings of the surfers are balanced, the actual numbers don't change. We say that [400, 200, 400] is a *steady state vector*, and from the previous discussion, we can see that, if we started at [1,000, 0, 0], our rank vectors will converge to this vector.

We might wonder whether this occurred because we started with the vector [1000, 0, 0]. Suppose we started with 250 people at A and C, and 500 people at B. After ringing the bell a few times, we'd again converge to the same steady state vector. As it turns out, no matter where our 1,000 surfers start, we will always have convergence to the [400, 200, 400] vector: PageRank is a *stable measure*.

The actual web requires a more nuanced approach because surfers aren't constrained to follow links. Page included a *teleport probability* in which a surfer would go to a random web page instead of a linked one. In the patent, Page suggests using a teleport probability of $p = 0.1$, so that 10% of the surfers on any given page will go to another page at random (including, possibly, the page they are on), while 90% follow the links. In this case, the rank vectors will converge to [392, 210, 398].

The teleport probability is especially important for pages linked like D, E, and F in Simplified Internet 2 of Figure 2.1. If surfers follow a link to D, they will have no way to leave it, so everyone would eventually end up on page D, and our rank vector would converge to [1,000, 0, 0]. On the other hand, if they can teleport, it's possible for them to escape. Therefore, if we use a teleport probability of $p = 0.1$, then the rank vector will converge to [933, 33, 33] after rounding.

If the number of web pages is small enough, we can find the steady state vector directly and exactly using algebra. However, this becomes infeasible if we're dealing with a large number of web pages. In practice, it's faster to find the rank vector by simulating the internet traffic using a discrete time model and an appropriate teleport probability; after a few hundred applications of the transition function, the rank vector is a reasonably good approximation to the steady state vector.

What does this rank vector tell us? Consider steady state vector [392, 210, 398], which is what our rank vectors for the Simplified Internet 1 would converge to for a teleport probability of $p = 0.1$, and remember that the web pages were selected because they were deemed relevant to a particular query. The rank vector suggests that, if 1,000 people begin navigating around this set of relevant web pages, then at some point, 38.5% can be found on page C, 35.6% can be found on page A, and 25.6% can be found on page B. Page C is the most popular and, under the assumption that the surfers know what they're looking for, page C should be ranked highest.

Derivative Work

At the time, Page and Brin viewed the Google project as an interesting distraction from their PhD studies. They approached several companies, including Yahoo! and Excite!, both founded by Stanford students, to see whether anyone was interested in buying their work. There was some interest, but the asking price — $750,000 — was too steep. Steven T. Kirsch, of InfoSeek, also turned down an offer to acquire Google.

By 2016, Google's ranking algorithm, known as PageRank (after its inventor and not its subject) allowed it to dominate the search engine market and become a company worth half a trillion dollars. Pundits said that Google won the search engine wars, which brings to mind a vision of engineers dumping other search algorithms into the trash bin.

The real picture is more complex. More often than not, technology advances when old ideas are joined to new ones. The older search methods, such as Boolean searches, thesaurus vectors, and so on, all form essential components of the newer search methods. Bharat and Henzinger went on to accept positions at Google; Yahoo! acquired AltaVista, and its engineers continue to develop improvements to their search engine technology. And while Yahoo! would be acquired by Verizon Wireless in 2016, its search engine patents were transferred to a new company, Excalibur, immediately before the sale. Every algorithm is a tool for solving some problem; the challenge lies in finding the right problem.

The current problem is *search engine spamming*, also known as *spamdexing*, which we can define as efforts to raise the ranking of a website independent of its contents. For example, in the early days of the internet, when Boolean searches and cosine similarity were used to rank pages, many websites filled their home pages with commonly used keywords ("car" or "stock market," repeated hundreds of times on a page). This would boost the frequency of the terms and the rank of a page would rise as a search result.

The pornography industry was notorious for this practice. A search for "cars" would invariably return "XXX GIRLS GIRLS GIRLS" among the top hits. While most users would ignore the obviously irrelevant pornography site, enough would visit it to make the practice worthwhile; at one point, everyone using a search engine expected to find several pornography sites among the search results for *any* term, no matter how innocuous.

When link analysis methods became the basis for search engines, spamdexers developed new strategies. In a *link farm*, a set of web pages is produced that link to the promoted page, increasing the in-degree of the page, the authority value of the page, and the number of surfers arriving at the page. In a *web ring*, the web pages have links to one another, reducing the number of surfers that go to other sites. For a while, the spammers seemed to have gained the upper hand, and Jeremy Zawodny (then at Yahoo!) went so far as to declare PageRank dead on March 24, 2003 (Figure 2.2).

However, if spamdexers can exploit weaknesses in PageRank, mathematicians and computer scientists can redesign PageRank to eliminate such weaknesses. One way to identify link farms and web rings appears in US Patent 7,509,344 for "Method of Detecting Link Spam in

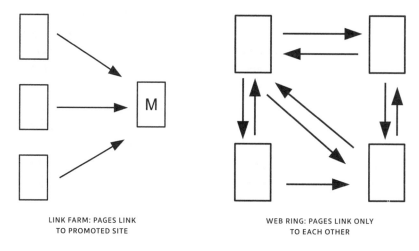

LINK FARM: PAGES LINK
TO PROMOTED SITE

WEB RING: PAGES LINK ONLY
TO EACH OTHER

Figure 2.2. Link Farms and Web Rings

Hyperlinked Databases," based on work done by Sepandar Kamvar, Taver Haveliwala, and Glen Jeh and assigned to Google on March 24, 2009. Their work relies on the *coupling probability*, the fraction of users who follow links. If the teleport probability is p, then the coupling probability will be $c = 1 - p$.

It should be obvious that changing the coupling probability will change the rank vector. However, because a large change in the coupling probability will presumably cause a large change in the rank vector, we want some standardized description of this change. We can look at the ratio between the change in the rank vector and the change in the coupling probability. Mathematically, we say we are looking at the *derivative of the rank vector with respect to the coupling probability*.

The derivative is a central concept of calculus, where students learn how to compute the derivative from a formula. To gain some insight into the meaning of the derivative, we'll compute it numerically as follows. Consider our Simplified Internet 1 of Figure 2.1. If we use coupling probability $c = 0.9$ (which corresponds to teleport probability $p = 0.1$), Simplified Internet 1 would have rank vector [391.9017, 209.6891, 398.4093].

Now suppose we increase the coupling probability slightly, from $c = 0.9$ to $c = 0.91$, corresponding to an increase of 0.01. If we do this, our rank vector becomes [392.7196, 208.6874, 398.5930]. The value for A will have increased by 0.8179; the value for B will have *decreased* by 1.0017; the value for C will have increased by 0.1837. Because c increased by 0.01, the *rate of change* for the ranks of A, B, and C will be 0.8179/0.01 = 81.79; –1.0017/0.01 = –100.17; and 0.1837/0.01 = 18.37. We say these values are the *average rate of change* of the ranks between $c = 0.9$ and $c = 0.91$. As long as the change in c is small, the average rate of change will approximate the derivative.

Of course, only politicians and pundits generalize from a single example, so let's see what happens if we take a smaller change, say, from $c = 0.9$ to $c = 0.901$. If we compute a new rank vector using this coupling probability, and compare it to the rank vector using $c = 0.9$, we find the rates of change for A, B, and C are 81.83, –100.52, and 18.59. An even smaller change, from $c = 0.9$ to $c = 0.9001$, corresponding to an increase of 0.001, would give us a rate of change of 81.51, –100.72, and 18.21. Notice that even though our change was smaller, the rate of change was more or less the same. Formally, we say the average rates of change converge to the derivative, which in this case will be around 82, –101, and 18.

Since the derivative is the ratio between the amount the rank vector changes and the amount the coupling probability changes, then the magnitude and sign of the derivative give us insight into how web traffic alters. It's convenient to interpret this derivative in terms of what happens when the coupling probability increases. In particular, what happens when more users follow links?

First, since the derivative of the rank vector is large and positive for A, then if more surfers follow links, more will arrive at A. Likewise, since the derivative is large but negative for B, then if more surfers follow links, *fewer* will arrive at B. Finally, the derivative is positive but more modest for C, so an increase in the coupling probability will lead to a modest increase for C. Intuitively, users gravitate toward web page A by following links, while they end at web page B more or less by accident. Web page C is the most interesting: in some sense, its popularity is less dependent on the structure of the internet.

How can we use this insight to identify link farms and web rings? In a link farm, surfers arrive at feeder sites mostly by chance. An increase in the coupling probability, which corresponds to a decrease in the teleportation probability, will produce a sharp decrease in the ranking of the feeder sites. And since the ranking of a page depends on the ranking of the pages that link to it, this would have the effect of decreasing the ranking of the target site significantly. Thus, the derivative for the target page of a link farm will be very large and negative.

How does this differ from a legitimately popular site? If the site is legitimately popular, then there are likely to be a few high-ranking sites that feed to it. While fewer surfers would teleport onto it, more would come to it following links from the high-ranking sites. These two effects mitigate each other, and while we might not be able to predict whether the site gains or loses rank, we can at least predict the change in rank will be modest. This suggests we can identify the target sites of link farms by focusing on sites whose ranks have large, negative derivatives with respective to the coupling probability.

What about web rings? As before, an increase in the coupling probability reduces traffic *to* the sites in the web ring. But an increase in coupling probability translates into a decrease in the teleport probability — so fewer visitors *leave* the web ring. All of the sites in the web ring will see their rank increase, and since the rank of a page depends on the rank of all the pages that link to it, the pages of the web ring will all have large, positive derivatives with respect to the coupling probability.

Again, a legitimate set of linked web pages would all see their ranks increased by the coupling probability. However, since these pages will also tend to have links to pages not in the web ring, some of that increase will be dissipated among other pages on the web, so the increase in rank will not be as great as it would be for the pages of a web ring.

In either case, we see that legitimate pages can be distinguished from link farms and web rings by the magnitude of the derivative, positive *or* negative. This suggests a search engine could limit the results it displays to those whose derivatives fall within a certain range: not too big and not too small.

Informational Entropy

Google's approach relies on finding the derivative of the rank vector, which in turn relies on the computationally intensive task of manipulating very large sets of data. A less intensive approach was developed by Yahoo! engineers Gilbert Leung, Lei Duan, Dmitri Pavlovski, Su Han Chan, and Kostas Tsioutsiouliklis; Yahoo! received US Patent 7,974,970 on July 5, 2011 for "Detection of Undesirable Web Pages."*

*Leung and Pavlovski left Yahoo! shortly thereafter and now work for Facebook and Google, respectively; the others are still at Yahoo!, while the patent itself is now held by Excalibur.

The Yahoo! engineers based their approach on the mathematical *theory of information*, developed by Claude Shannon (1916–2001), who worked at Bell Labs on wartime projects during the 1940s. Shannon's work was classified until 1948, when his groundbreaking paper "A Mathematical Theory of Communication" appeared.

Suppose you have an oracle that will send you a letter that includes the answer to any question you ask. It might seem that the letter contains information. But communication is a two-way street: if you throw the letter away without reading it, then the letter contains no information.

You might wonder why we'd discard a letter without reading it. But most of us routinely delete emails with subject lines like "You May Have Won $1,000,000!!!!" or "Nigerian Prince Seeks Business Partner." Shannon's basic insight consists of two parts.

To begin with, consider the email with subject line "You May Have Won $1,000,000!!!!" The information contained in this letter answers a single yes/no question, namely, "Have I won $1,000,000?" But most of us know in advance that the answer to this question is almost certainly no. Since we already know the answer contained in the email, the email itself contains no information, and we can delete it without reading it.

What might not be obvious is that the same is true if the answer to a question is almost certainly yes. If you've taken out a car loan, an email with the subject line: "Your next bill is available online" answers the question, "Do I have to pay my installment loan this month?" Again, we already know the answer to the question, so we can delete the email: it contains no information.

Putting these together, we might conclude that the information content of a message rests not on the answer it contains but rather on how much it reduces the uncertainty of the answer. If the message is certain to contain the answer yes, then there is no reduction in uncertainty; the same applies if the message is certain to contain the answer no. In either case, the message contains no information. It's only when there's some uncertainty over the answer that the message contains information.

Suppose the information can be found by the answers to a sequence of yes/no questions. We can record each yes answer as a 1 and each no answer as a 0, so the information can be represented as a sequence of 1s and 0s. Since a single 1 or a single 0 corresponds to the answer to a single yes/no question, and thus to the smallest possible amount of information, Shannon's coworker, John Tukey (1915-2000), suggested the name *bit*, a contraction of *binary digit*, for this smallest particle of information.

Surprisingly, *any* information can be found by asking a sequence of yes/no questions; this is the basis behind the game Twenty Questions. But it can sometimes be difficult to determine what those yes/no questions must be, which is why Twenty Questions is a game: a poor player might have to ask many questions, while an expert player might only need a few. Put another way: the letter sent by the oracle would be a sequence of 1s and 0s, but the answer would only make sense if we knew what our yes/no questions were.

To avoid the problem of determining which questions should be asked, Shannon introduced *informational entropy*. Suppose the question could have one of several responses A, B, C, and so on. If $1/a$ of the time, the oracle answered A, $1/b$ of the time the oracle answered B, $1/c$ of the time the oracle answered C, and so on, then the amount of information conveyed by a letter sent by the oracle will be

$$(1/a) \log a + (1/b) \log b + (1/c) \log c + \ldots$$

where, as before, we define $\log N$ as the number of successive doubles we must undertake before arriving at N. For example, if the answer to our question was always no, then $1/1$ of the message would be no, and the information delivered by the letter would be $(1/1) \log 1 = 0$. On the other hand, suppose half the time the answer was yes and half the time the answer was no. Then the letter would convey $1/2 \log 2 + 1/2 \log 2 = 1$ bit of information.

Or consider the question, "What color should we paint the front room?" If the answer was *always* blue, then a message containing the answer would convey 0 bits of information. On the other hand, if $1/2$ the time, the answer was blue, $1/4$ the time it was green, $1/8$ the time it was yellow, and $1/8$ the time it was red, then the information content of a message containing the answer would be

$$(1/2) \log 2 + (1/4) \log 4 + 1/8 \log 8 + 1/8 \log 8 = 1.75 \text{ bits}$$

How can this help defeat spamdexing? First, consider a link farm. The target site for the link farm has a distinctive feature: many of its incoming links (its *inlinks*) will be from the feeder sites. Since the target site can't rely on legitimate incoming links (because then there'd be no need to establish a link farm), these sites are probably hosted on the same web-connected computer (the *server*), controlled by the target site's owners.

Now consider the answer to the question, "Which server originated incoming traffic?" If all incoming traffic came from the same server, then the informational entropy of the answer would be 0. On the other hand, a legitimate site would have incoming traffic from several different servers, and the answer would have a higher informational entropy.

Similarly, we might ask the question, "Which server did outgoing traffic visit?" In a web ring, the outgoing traffic will generally be directed to other pages on the same server, so the answers will generally be the same, and the informational entropy of the answer will be low. In a legitimate set of connected web pages, at least some of the outgoing traffic will go to other servers, so the answers will be different, and the informational entropy of the answer will be higher.

As with the derivative of the ranks, we can distinguish legitimate pages from link farms and web rings by the informational entropy of the traffic pattern, and our search engine could limit the results it displays by only showing websites with a sufficiently high informational entropy.

The ~~End~~ Start of the Spam Wars

In 2015, consumers spent over a trillion dollars online. Many of these purchases began with an internet search. With so much money at stake, a new profession has emerged: the search engine optimizer (SEO). A SEO is a person or machine tasked with increasing a website's profile on search engines. They are the modern equivalents of advertising executives, but instead of appealing to human psychology to promote a product, they design websites that will be highly ranked by mathematical algorithms. Media analyst group Borrell Associates forecasts that SEO-related spending will top $80 billion by 2020.

Spamdexers have the same relationship to SEOs that con artists have to salespersons: both seek to sell a product by appealing to how we perceive value. They differ primarily in the perceived value of the product: one of them sells get-rich-quick schemes and worthless real estate, while the other sells investment opportunities and timeshares.

Currently, search engines have the upper hand: in business (swindling or advertising) they represent the resistant targets, those not susceptible to the techniques of the con artist and the salesperson. But mathematics can be used by anyone, and as long as the ranking algorithms for search engines are known, pages can be optimized to raise the rank of a website. Far from being over, the spam wars are just beginning.

3

\\\\\\\\\\\\\\\\\

A Picture *Is* a Thousand Words

Some say they see poetry in my paintings; I see only science.

GEORGES SEURAT (1859–1891)

In the 1830s, the Gobelins Manufactory in France, which produced tap-
estries, had a problem: black thread didn't always look black. Company
executives asked the chemist Michel Eugène Chevreul (1786–1889) to
look into the problem. Chevreul concluded the problem was the eyes,
not the dyes. In particular, our perception of a color is affected by the
colors of adjacent objects. This led Chevreul to the theory of *simultaneous contrast*, which he described in an 1839 publication.

Simultaneous contrast offered artists a new way to produce color.
Traditionally, pigment mixtures are based on *subtractive color*: for example, if you mix red and green paint, it appears grayish green. But
in 1851, German physiologist Hermann von Helmholtz (1821–1894) suggested the human eye can only detect three colors: red, green, and blue.
All other colors we see are the result of *additive color*: if we see red and
green simultaneously, we perceive yellow. English physicist Thomas
Young (1773–1829) made a similar suggestion in 1802, though Helmholtz's physiological mechanism is more nearly correct; consequently,
we speak of the Young-Helmholtz theory of color vision.

As the name suggests, additive color has the potential for producing brighter, more vibrant colors. Georges Seurat (1859–1891) used this technique in his *A Sunday Afternoon on the Island of La Grande Jatte*, first exhibited in 1886. The painting depicts a scene in a Paris park. To achieve the impression of color, Seurat placed small dots of pure colors, mostly red, green, blue, yellow, and white, in proximity. At a distance, the individual dots were indistinguishable, but the light reflected from the dots reached the eye and, through additive color, produced the desired color effect. Seurat's work was distinctive enough that it created a new style of painting, named after a derisive review of it: *Pointillism*.

Seurat's work leads to a natural way to produce, to analyze, and to compare images. First, an image can be divided into a number of blocks, now known as *pixels* (a word of uncertain derivation, but likely a contraction of *picture element*). Each pixel has a color that, like a pointillist painting, can be described in terms of a few primary colors. Seurat used red, green, blue, yellow, and white dots, though most modern electronic displays only use the first three, giving rise to RGB color. The amount of "paint" is determined by the intensity of each of the colors, given on a scale from 0 (no light) to 255 (maximum light). This allows us to give the color of any pixel using three numbers: the RGB intensity.

For example, consider a pixel with RGB intensity (0, 0, 192), which we read as indicating red intensity 0, green intensity 0, and blue intensity 192. We perceive this as a blue pixel. We can darken or lighten it by changing the intensity; thus, (0, 0, 191) is still blue, though slightly less bright. Equal mixtures of red, green, and blue produce "white," perceived as a shade of gray, depending on the intensity. Thus, (255, 255, 255) would be perceived as white, while (100, 100, 100) would be grayish, and (0, 0, 0), where all intensities are 0, would in fact be black.

For illustrative purposes, suppose we have two paintings, both of which consist of 4 × 4 = 16 pixels, and suppose that each pixel has 0, 1, 2, or 3 blue dots (hearkening back to Seurat and the work of the Pointillists). We can then represent each picture using a *blue color histogram*, which would show us the number of pixels with 0, 1, 2, or 3 dots. If our painting also had red and green dots, we could also produce the red and green color histograms (Figure 3.1).

Mathematically, a histogram (also known as a bar chart) is a vector, where each component records the number of pixels of a certain intensity. Our first painting has 3 pixels with 0 blue dots, 2 pixels with

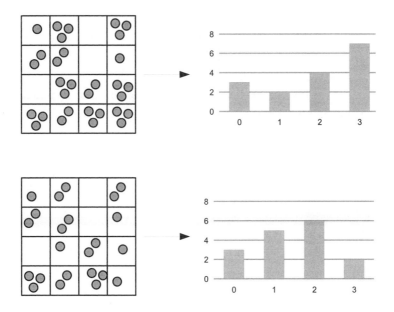

Figure 3.1. Color Histogram

1 blue dot, 4 pixels with 2 blue dots, and 7 pixels with 3 blue dots. We can describe the painting using the *feature vector* [3, 2, 4, 7], whose four components give the number of pixels with 0, 1, 2, or 3 blue dots, respectively. Meanwhile, our second painting has 3, 5, 6, and 2 pixels with 0, 1, 2, or 3 blue dots, respectively, and gives us the corresponding feature vector [3, 5, 6, 2].

Since the RGB intensity of each pixel can range from 0 to 255, our feature vector might have 256 components, corresponding to the number of pixels with blue intensity 0, blue intensity 1, blue intensity 2, and so on. In practice, we divide the range into a number of *bins*; for example, we might designate all pixels with intensity between 0 and 63 as having a "very low" intensity, all pixels with intensity between 64 and 127 as having a "low" intensity, all pixels with intensity between 128 and 191 as having "medium" intensity, and all pixels with intensity between 192 and 255 as having "high" intensity. We can then record the color histogram as a four-component vector, where the components correspond to the number of pixels with very low, low, medium, or high intensity. A car in a picture might have blue color histogram [180, 420, 1550, 350], from which we can determine the image itself took up 180 + 420 + 1,550 + 350 = 2,500 pixels.

It's worth comparing the RGB triplet and color histogram to the bag of words model. Both rely on vectors: the color histogram, recording the number of pixels of a given color, and the document vector, recording the number of times a given token appears. Both ignore order: the location of the blue pixels in an image or the tokens in a document is not included in the vectors. To a mathematician, a picture isn't *worth* a thousand words, it *is* a thousand words.

This raises an important problem with patents. Patents are issued for a *device*, and don't depend on how the device is used: If you receive a patent for a hammer, used to drive nails, another inventor *can't* receive a patent for the same hammer, used to crack walnuts. But this means that, in principle, a patent for an invention that compares two *documents* could be construed as covering the use of the invention to compare two *pictures*.

Product (Mis)placement

Currently, the patent office draws a distinction between such uses, so it's possible to obtain a patent for a device comparing two pictures, even if the underlying mathematics is similar or identical to the mathematics used to compare two documents.*

One important commercial application concerns the appearance of a manufacturer's product. For example, in the 2012 movie *Flight*, released by Paramount Pictures, Denzel Washington plays an alcoholic pilot. In one scene, he opens up a hotel minifridge, and several common brands of beer and alcohol are clearly recognizable. Anheuser-Busch, the maker of Budweiser, complained to Paramount and asked that the image of their beers be removed, out of fear that people might associate their products with alcohol abuse.

More generally, a company might want to know how and where its products appear in the media. Or a company might pay to have an advertisement put on a bus and want detailed information about how often the bus passes a particular corner and how visible its advertisement will be. In either case, finding when, where, and how the image appears could require reviewing thousands of hours of video, so an

*More important, no one has tested the limits of such claims. One risk is that, since the underlying mathematics is old, *all* such patent claims might be deemed invalid.

automated technique of identifying images in a picture or video would be a valuable addition to the tools of the market researcher.

It's not too surprising that Nielsen, the global marketing giant, holds several patents on the use of histograms to identify images. One of these is US Patent 8,897,553 for "Image Comparison Using Color Histogram," invented by Morris Lee, Nielsen's principal researcher, and granted on November 25, 2014.

Suppose we have a reference image, say, the picture of a can of beer. It's easy enough to produce the color histogram for this image. A movie like *Flight* consists of perhaps 200,000 pictures, shown in rapid succession to provide the illusion of movement. Sophisticated algorithms can be used to separate each picture into a collection of images of individual objects. A tableau, consisting of a refrigerator shelf with several cans of beer and some jars of pickles, could be separated into individual images of each can of beer and jar of pickles. These images could then be analyzed to identify their contents.

There are a number of ways we can make the comparison, but the Nielsen patent is based on the *histogram difference* between the reference and query images. First, we need to scale the query image so that it is the same size as the reference image. For example, if our reference image has blue color histogram [180, 420, 1,550, 350], and our query image has blue color histogram [40, 80, 300, 80], then the reference image contains 180 + 420 + 1,550 + 350 = 2,500 pixels, while the query image contains 40 + 80 + 300 + 80 = 500 pixels. This means we'll need to scale the query image by a factor of 5, turning it into [200, 400, 1,500, 400].

How can we compare the images? In 1991, Michael Swain, at the University of Chicago, and Dana Ballard, at the University of Rochester, suggested using *histogram intersection* as a means of comparing two images. To find the histogram intersection, we look at the corresponding components of the two histograms, and choose the lesser value: this is called the *componentwise minimum*. We then sum the componentwise minima. In this case, our two vectors are [180, 420, 1,550, 350] and [200, 400, 1,500, 400]. The first components of the two are 180 and 200; the minimum is 180. The second components are 420 and 400; the minimum is 400. The third components are 1,550 and 1,500; the minimum is 1,500. And the fourth components are 350 and 400; the minimum is 350. Our histogram intersection will be the sum of these numbers: 180 + 400 + 1,500 + 350 = 2,430.

What does the histogram intersection mean? If we continue to interpret the four components of our blue color histogram as the number of pixels with blue intensity very low, low, medium, and high, then we can interpret the componentwise minima as follows. The first components are 180 and 200, which means the first picture has 180 pixels with a very low blue intensity, while the second has 200 pixels with a very low blue intensity. At least 180 pixels in both images have very low blue intensities. Similarly, at least 400 pixels in both images have low blue intensities; at least 1,500 pixels in both images have medium blue intensities; and *at least* 350 pixels in both images have high blue intensities. Altogether, at least 2,430 pixels in both images have the same blue intensities. Since each image consists of 2,500 pixels, this means that 2,430/2,500 = 97.2% of the pixels in both images have the same blue intensities. In some sense, the two objects have a very similar (97.2%) level of blueness. We can repeat the process for the other color histograms. If all three are sufficiently similar, then we might deem the query image to be another picture of the object in our reference image.

Nielsen's patent actually uses a variation, the histogram difference: this is the complement of the intersection, so if two images have an intersection of 97.2%, then they have a difference of 2.8%. Since we can form a histogram for each of red, blue, and green from the RGB triplets, we can find three histogram differences. A sizable difference in *any* of the histograms would be sufficient to declare the object in the picture to be unlike the object in the reference library.

The Geometry of Color

Pictures and documents share another similarity: both are subject to the problems of polysemy and synonymy.

In the case of the color histogram, the problem originates with the binning process. Suppose our image consists of a single pixel with RGB triplet (0, 0, 192), which we'd perceive as blue. Using the same four bins as before, the blue color histogram for this image will be [0, 0, 0, 1].

Now consider a second image, also consisting of a single pixel, this time with RGB triplet (0, 0, 191). While the intensity is less, the difference is so small that few viewers would see it, and we'd regard this as a blue pixel. But because of the way we've set up our bins, our blue color histogram is now [0, 1, 0, 0], and by histogram intersection, the two images will have 0 pixels with the same intensity: they are 0% alike (and so the histogram difference is 100%). Yet both pictures consist of a single blue dot of virtually identical intensities.

The problem is that RGB triplets mix information about *luma* (intensity) and *chroma* (color). Any pixels with RGB intensity $(0, 0, b)$ will appear blue, regardless of the actual value of b; the only difference is that the higher the value of b, the brighter the pixel. If we can separate luma from chroma, then we might be able to improve our ability to distinguish between images. Mathematically, we are looking for a *transformation of coordinates* that allows us to go from one *color space* into another. More generally, we want to find a *geometry of color*.

It might seem peculiar to talk about the geometry of color, since we ordinarily associate geometry with shapes. For millennia, mathematicians thought the same way. But at his inaugural lecture at the University of Erlangen, in Bavaria, the German mathematician Felix Klein (1849–1925) identified geometry as the study of the properties of objects that are *invariant* under a specified set of allowable *transformations*.

For example, two transformations we can apply to real objects are *translations*, which move objects from one place to another, and *rotations*, which turn an object around a point. Properties such as volume, areas, and angles remain unchanged under these transformations, and so geometry studies these properties.

Two common geometric transformations can't be done with real objects: we can produce the mirror image of a figure (a *reflection*), and we can scale the figure to make it larger or smaller (a *dilatation* or, more commonly, a *dilation*). Transformations are usually classified into two categories. An *isometry* (from the Greek words meaning "same measurements") leaves lengths unchanged: translations, rotations, and reflections are isometries. In contrast, a *conformal transformation* leaves angles unchanged: a dilation is best known as a conformal transformation. Note that every isometry is also a conformal transformation. Euclidean geometry, the type most of us study in school, can be described as the study of conformal transformations.

Consider the color histogram of the image of an object. This color histogram will be invariant under translations, rotations, and reflections, and if we scale the histogram, it will also be invariant under dilation. The color histogram is an invariant under the set of conformal transformations and is a geometric object in its own right.

However, there are other possible transformations. For example, we might raise or lower the intensities of the three colors. We might shine more light on the object, or we might view the object through a filter. Since the real color of an object is determined by its physical composition, we might look for a way to describe color that is invariant under such transformations. One approach is used in US Patent 8,767,084, invented by Ravi K. Sharma, and assigned to Digimarc Corporation, a leading provider of image recognition technology, on July 1, 2014.*

One of the ubiquitous features of modern society is the UPC (Universal Product Code), which is rendered as a set of thin and thick bars. A scanner can read the UPC and information about the product, notably its identity and price, can be found immediately and added to a bill. But anything that prevents the scanner from reading the UPC barcode, such as the placement of the object as it passes over the scanner, or a smudge on the barcode, will make the UPC barcode unreadable.

As an alternative, we can rely on image recognition technology. A scanner might take a photograph of an object, produce the color histogram, and then compare it to the color histograms in a library of images to identify the product. The possibilities go far beyond the cash register: a person might take a photograph of an object, and an application could identify what it is and direct the user to a website where the item could be purchased.

The Digimarc patent uses a comparison of color histograms based on the L-RG-BY (luminance, red-green, blue-yellow) triplet. These are computed using the formulas

$$D = \text{Maximum of RGB} - \text{Minimum of RGB}$$
$$L = (R + B + G)/3$$
$$RG = (R - G)/D$$
$$BY = (B - (R + G)/2)/D$$

For example, consider a pixel with RGB triplet $(150, 100, 200)$. The maximum value of RGB is 200, while the minimum is 100, so D is $200 - 100 = 100$. Luminance is $(150 + 100 + 200)/3 = 150$. RG will be RG of $(150 - 100)/100 = 0.5$, and BY will be $(200 - (150 + 100)/2)/100 = 0.75$.

What do these numbers mean? Luminance is the simplest: it is

---

*Sharma worked as a software engineer for both DEC and Intel before joining Digimarc in 1998.

the average intensity of the three colors. The other values are somewhat harder to interpret. To understand the basis for them, consider the problem of identifying compass directions. There are the four *cardinal directions*: north, south, east, and west. If you face one of these directions and turn partway toward an adjacent direction, you might describe your facing in terms of the original cardinal direction, and the direction you've turned toward. If you begin by facing north, turn toward the west ("to your left"), if you've turned far enough, you'll be facing west. But if you stop midway through the turn, you'll be facing northwest. In this way, we can produce four *intercardinal directions*: northwest, northeast, southwest, and southeast. We can view the intercardinal directions as a primary direction modified by a secondary direction: to face northwest, begin by facing north and then turn halfway toward the west.

We can use this approach to identify colors, where we use the primary colors in place of the cardinal directions. In the L-RG-BY triplet, the color is identified by RG and BY, and in this case, we might describe our color as "0.5 RG by 0.75 BY." As before, we can construct the color histograms, though in this case, we'll only have two: the RG color histogram, and the BY color histogram.

One advantage to L-RG-BY space appears as follows. Suppose we shine more light on an object. To avoid having the light itself cause a color change, we'd use white light, which corresponds to increasing the R, G, and B intensities by the same amount. For example, if we increased each of these intensities by 50 units, the pixel with RGB triplet $(150, 100, 200)$ becomes pixel with RGB triplet $(200, 150, 250)$. This will change the value of L by 50 units, which reflects the increase in intensity. But the value of D will be unchanged, since both maximum and minimum have increased by the same amount.

What about RG and BY? Since both R and G increase by 50, R − G remains the same and, since D also remains the same, RG will be unchanged. By a similar argument, $(R + G)/2$ will increase by 50, while B increased by 50, so $B − (R + G)/2$ will remain unchanged, and so BY will also remain unchanged. Our color, which was "0.5 RG by 0.75 BY," is still "0.5 RG by 0.75 BY."

Another advantage to L-RG-BY color space is that the color histograms are also invariant under a proportional change of the color intensities. This might occur if we view the image through a filter. For example, if our filter cut the intensities of R, G, and B by half, then the RGB intensity of a pixel might change from (150, 100, 200) to (75, 50, 100). Since all the intensities have been cut in half, D will also be cut in half: it will go from 200 − 100 = 100 to 100 − 50 = 50. When we compute RG and BY, we find that this decrease in D compensates for the changes in R, G, and B: thus, RG will be (75 − 50)/50 = 0.5 and BY will be (100 − (75 + 50)/2)/50 = 0.75, which are the same values as before. Once again, the change in intensity does not change our color: it is still "0.5 RG by 0.75 BY."

Points of Similarity

Color histograms work best when the query and reference images have some standardized orientation, and different objects have radically different color histograms. They're good for comparing the product in an image with a product for sale, since two products are either different enough to make the histogram comparison effective or so similar that images are not enough to distinguish between them. You wouldn't confuse a sports car with a minivan, but there's so little physical difference between laptops that comparing their pictures is almost meaningless.

But what if there aren't standardized photographs or the distinctions between images are subtle? For example, consider the photograph of a building. Many buildings look alike, and the color of a building can change due to ambient lighting conditions: a building will appear one color in the bright sunlight of a midsummer noon and a different color on an overcast mid-December morning. Thus, histogram comparison methods won't be effective.

The situation is even worse because buildings are objects that exist in three dimensions, while photographs can record only two. This means that *where* the photograph was taken can radically alter the appearance of the building. So how can we compensate for these distortions? As before, we look for invariants, which means we need to turn to geometry. In this case, we need to turn to a *projective geometry* (Figure 3.2).

Since the Renaissance, artists understood that one way to produce a realistic image of an object was to use perspective. Suppose we want to paint a pastoral scene showing some sheep, a shepherd, a few background trees, and the distant mountains (Figure 3.2). If the picture faithfully reproduces how the object appears, then the line from

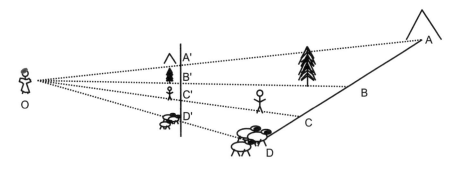

Figure 3.2. Projection onto Canvas (Sans Artistry)

your eye (O) to the image of the object on the canvas (A', B', C', or D') should, if extended, meet the corresponding part of the object in the real world. Mathematically, we are *projecting the objects onto the plane of the canvas.*

Projections are not isometries: the distance between the mountain and the tree on the canvas is nowhere near the distance between the two in the real world. Nor are projections conformal: railway tracks in a picture appear to converge, though they are in fact parallel. What makes projective geometry a *geometry* is that certain ratios are invariant under projection. The first of these invariant ratios was discovered by French mathematician Girard Desargues (1591–1661), who is generally credited with being the founder of projective geometry.

Consider the line between the mountain and the flock. If we extend the lines of sight to the tree and the shepherd, they will meet this line, and the four real objects will correspond to the points A, B, C, and D in the real world; meanwhile, the images of the four objects will correspond to the points A', B', C', D' on the canvas. Note that only the mountain and flock are actually at these points; the alert reader will recognize that B and C are the projections of the tree and shepherd onto the line AD. Desargues discovered that whatever the actual lengths of AC, BC, AD, and BD, and whatever the actual lengths of A'C', B'C', A'D', and B'D', the *cross ratios* will be equal: $\frac{AC/BC}{AD/BD} = \frac{A'C'/B'C'}{A'D'/B'D'}$. This result is known as Desargues's theorem.

For example, suppose AC = 8, BC = 4, and AD = 10, BD = 6, all in miles, while A'C' = 316, B'C' = 203, and A'D' = 492, B'D' = 378, all in pixels. Then $\frac{8/4}{10/6} = \frac{316/203}{492/378}$ (to within the roundoff error).

What makes this useful is suppose we take the same points A, B, C, D (*not* necessarily the same objects) and paint a picture from a different location. The cross ratio will be the same. We might find A′C′ = 434, B′C′ = 90, and A′D′ = 458, B′D′ = 114, pixels. Once again, we can verify that $\frac{8/4}{10/6}=\frac{434/90}{458/114}$. This suggests a way we can determine whether two pictures are different views of the same scene.

Unfortunately, it's not feasible to translate Desargues's theorem into a useful way to compare two images. The most obvious feature of an image is the objects in a picture. While Desargues's theorem uses the distances between these objects in the image, it requires the distance between the projection of these objects onto a distant line. This in turn requires knowing where the picture was taken from.

To circumvent this difficulty, we can proceed as follows. The cross ratio is the quotient of two fractions: AC/BC divided by AD/BD. The projection will change the lengths, AC to A′C′, BC to B′C′, and so on. However, Desargues's theorem is that the new quotient will be the same as the original.

We might ask: How can we alter the terms of a quotient without changing the quotient itself? In general, if we multiply both terms of a quotient by the same amount, we'll leave it unchanged: thus, 12 divided by 3 is the same as 12 × 5 = 60 divided by 3 × 5 = 15. This means that one way we can keep the cross ratio invariant is to multiply the two fractions, AC/BC and AD/BD, by the same amount.

What happens when we multiply a fraction by an amount? Suppose we multiply AC/BC by p/q. Then we'll multiply AC by p and BC by q. If we multiply AD/BD by the same fraction, we'll again multiply AD by p and BD by q. What this means is that if AC, BC, AD, BD, and A′C′, B′C′, A′D′, B′D′ are lengths in two projections of the same set of objects, then we'll get the second lengths from the first by multiplying by one of two numbers: *either p or q.*

We can use this as follows. Suppose we have two images and these lengths. Consider the ratio between the lengths A′C′ and AC; this corresponds to the value of A′C′/AC. This would give the amount we need to multiply the length AC by in order to obtain the length A′C′. Similarly, the ratio between the lengths B′C′ and BC, which corresponds to B′C′/BC, would be the amount we'd need to multiply BC by to obtain B′C′; and likewise for the remaining length ratios.

There's no a priori reason to suspect these ratios have anything to do with each other. We might need to multiply AC, BC, AD, and BD by 2, 3, 1/2, and 3/4 to obtain A'C', B'C', A'D', and B'D', so obtaining the second set of lengths from the first set requires multiplying by one of four different numbers. But *if* the lengths AC, BC, AD, BD and A'C', B'C', A'D', B'D' come from two projections of the same set of objects, *then* we can get the second lengths from the first by multiplying by one of *two* numbers. This suggests the following. Suppose we have two images that are projections of the same object. If we find the ratios between corresponding lengths in the two images, these ratios will have just a few distinct values. On the other hand, if the two images are *not* projections of the same object, then the ratios will be more scattershot.

In 2010, a team consisting of Sam Tsai, David Chen, Gabriel Takacs, Vijay Chandrasekhar, Bernd Girod, from Stanford University, and Ramakrishna Vedantham and Radek Grzeszczuk, from Nokia, developed this approach. Their initial work suggested that it would be very effective in distinguishing between images. However, they only developed a mathematical algorithm for identifying images. In order to be patentable, a mathematical algorithm needs to be joined to some machine or process. This would be done by Giovanni Cordara, Gianluca Francini, Skjalg Lepsoy, and Pedro Porto Buarque de Gusmao; Telecom Italia would receive US Patent 9,008,424, granted on April 14, 2015, for their work.

Imagine that a reference and query images have a number of key points defined: for example, in the picture of a building, the key points might be the location of the main entrance, windows, and so on. Given any two key points, we can define a distance between them: for example, the distance between the main entrance and an adjacent ground floor window. We can then form the *log distance ratio* (LDR): the log of the ratio of the distance between the key points in the query image to the distance between the corresponding key points in the reference image. Since there are many distances that could be compared, we can form the LDR histogram for all the ratios, and compare the histograms to decide whether the images are similar. If two images are projections of the same object, we'd expect less variability in the LDRs than if the two images were projections of different objects (Figure 3.3).

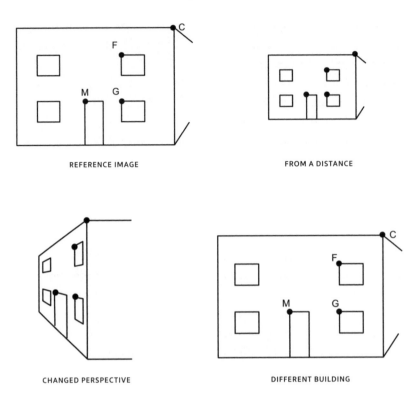

Figure 3.3. Building Images: Zoom, Perspective Change, Different Structure

For example, suppose we have four key points in our reference image: M, a point on the front door frame; G, a point on a ground floor window frame; F, a point on an upper floor window frame; and C, a point on the corner of the building. We can find the distance, in pixels, between our key points. As there are four points, there will be six distances. Suppose we determine MG = 250, GF = 250, FC = 269, MF = 354, GC = 430, and MC = 610.

Now suppose we take a picture of the building from farther away; equivalently, we might digitally zoom out. The distances will change proportionally: we might find MG for the new image is 125, GF is 125, and so on.

To find the LDRs, we find the logs of the ratio of a distance in the query image to a corresponding distance in the reference image. The

LDR for MG would be $\log(125/250) = \log(1/2) = -1.$* Since all the distances changed by the same amounts, the LDR for all the distances will be the same.

What if we construct the histogram for the LDRs? Since all the LDRs have value -1, then all three will end in the same bin. Our histogram will have a very unusual appearance: it will consist of a single "spike," corresponding to the bin that includes the LDR of -1, with all the other bins left empty. Moreover, since the distances are invariant under rotations, reflections, and translations, then the LDR histogram will in fact be invariant under any conformal transformation. Finally, unlike color histograms, the LDR histograms won't be affected by changes in lighting or color.

Now suppose we take a picture of the same building from a different location. As before, we'll measure the distances between the key points in pixels, then compute the LDRs. Suppose we do this and find MG = 18, GF = 133, FC = 123, MF = 121, GC = 255, and MC = 243.

What happens when we compute the LDRs? For MG, the LDR will be $\log(18/250) \approx -3.729$, for GF it will be $(133/250) \approx -0.907$, for FC it will be -1.13, for MF it will be -1.55, for GC it will be -0.756, and for MC it will be -1.327. While the LDRs don't have the same values, they fall in a relatively narrow range, between -0.756 and -3.729. Moreover, depending on how we choose our bins, there may be a distinct spike in the LDR histogram, because there are distinct clusters of values: for example, GF, FC, and GC are relatively close to one another, as are MF and FC.

In contrast, suppose we have a different building: perhaps one of the same height and width, but where the windows are slightly farther from the door. As before, we measure the distances: suppose MG = 350, GF = 250, FC = 180, MF = 430, GC = 381, and MC = 610. As before, we'll compute the LDRs. For MG, the LDR will be $\log(18/350) \approx -4.214$; for GF, the LDR will be -0.907; for FC, -0.552; for MF, -1.833; for GC, -0.580; and for MC, -1.327. This time there is more variability in the LDRs, from -0.552 to -4.214, which suggests that the image might not be that of this other building.

---

*If the log is negative, it corresponds to the number of times we must *halve* 1 to get an amount. Thus, to get 1/8, we must start at 1 and halve three times: $1 \to 1/2 \to 1/4 \to 1/8$, so $\log(1/8) = -3$.

4

〰〰〰〰〰〰〰

If You Like Piña Coladas

A surgeon, in practice, is desirous of meeting with a suitable partner.
He is fair, 47 years of age, a bachelor, fond of domestic
and farming pursuits, good tempered, and likes children. . . .
Would like to hear from Nos. 6731, 6732, 6693 . . .

FROM *THE MATRIMONIAL NEWS*, AS QUOTED IN *CHAMBERS JOURNAL*
Saturday, June 7, 1873

On April 12, 1873, Eleanor Berry, a 22-year-old schoolteacher in Gilroy, California, responded to a personal advertisement by Louis Dreibelbis, a miner in Grass Valley, California. After a three-month correspondence, she agreed to marry him and boarded a stagecoach to Grass Valley. Also on the stagecoach was a strongbox containing $7,000 in gold, destined for a Grass Valley bank. En route, bandits stopped the stagecoach and demanded the strongbox. They picked the outer lock and were about to blow the inner lock with gunpowder. As the strongbox was next to her own luggage, Eleanor requested they move the box off the stagecoach first, and they complied. After they took the gold, the bandits departed, and the stagecoach and passengers made their way to Grass Valley, where Eleanor met Louis, recognizing him as one of the bandits. Louis fled, and Eleanor returned home to Gilroy.

"Overweight 45-Year-Old Male Seeks Svelte 26-Year-Old Female"

Criminals posing as honest citizens aside, personal ads pose unique implementation challenges. Consider the seemingly similar problem of matching the seller of a vehicle to a buyer. The transaction is clearly defined: one person with a car seeks a person interested in buying a car. The differences between cars are small and can be expressed using just a few characteristics: make, model, color, year, and mileage, all of which can be listed in the ad. The buyer usually has a constraint (maximum amount they're willing or able to spend on the car) and a clear vision of what they want (green minivan), so they can scan the ads quickly and see if there are any matches.

Finding a potential date or mate is far more complicated. First, there are so many features that distinguish one person from another it is impossible to list all of them, even when they are reduced to incomprehensible acronyms: DWF ISO SWM AL GSOH WTR.* Moreover, the author invariably omits details that, in his or her own mind, are irrelevant ("garlic aficionado") or damning ("stagecoach bandit"). The biggest challenge is that "good date" or "ideal spouse" mean different things to different people: in mathematical terms, the dating problem is *ill defined.* Such problems are the bane of the information age, for they cannot be solved by computers.

To solve an ill-defined problem, we must make it well defined. Rather than trying to find a "good date" or "an ideal spouse," we might begin by finding a person with a specific set of characteristics. This is the basis for US Patent 6,272,467, developed by Peter Durand, Michael Low, and Melissa Stoller and assigned to Spark Networks Services in 2001. Spark Networks Services would be acquired by Advanced Telecom Services in 2005, which now operates a dating service under the name Match-Link, offered to radio and newspaper outlets as a way of earning NTR (non-traditional revenue). To avoid arduous phrases like "the male or female person looking for a mate or date," we'll say instead that a *client* is seeking a *match* and refer to both as *members* of a matchmaking service.

*"Divorced white female in search of single white male animal lover good sense of humor willing to relocate."

Imagine having each member complete a form with two parts. In the first part, the member describes his or her ideal match; in the second, the member describes his or her own characteristics, particularly those relevant to the characteristics of an ideal match. If age is a consideration in the suitability of a match, then the member would be prompted to enter his or her age. The database of member information could then be searched to find a match for any client. The problem of finding an ideal match is mathematically identical to the problem of comparing two documents for similarity.

Of course, you can't always get what you want: The inventors note, "Many users of dating services do not make rational choices for their ideal match given their own profile,"* and give as an example an overweight, middle-aged man who only wishes to date very thin women between 26 and 28 years of age. Thus, they included several adjustments to build and age criteria.

Among the problems the aforementioned client might encounter in his quest for a match is that the age range — 26 to 28 years of age — is too narrow. A commonly quoted rule of thumb is *half plus seven to double seven less*: You should date people whose ages are between half your own, plus seven years; and twice seven years less than your age. A 40-year-old should date people between the ages of $(40 \div 2) + 7 = 27$ and $(40 - 7) \times 2 = 66$.

Based on their research, the inventors concluded that the desired ages of a match should span at least 6 years for men under 40, and at least 10 for those 40 and older. At the very least, the age range must be expanded. On the assumption that heterosexual men are favorably disposed toward dating younger women, the software automatically adjusts the minimum age for males according to the formula

Minimum Age = Desired Minimum − (Desired Minimum/10) − 1

A 40-year-old might *indicate* a desire to date women between the ages of 26 and 28, but the software would adjust the minimum age to 26 − (26/10) − 1 = 22.4 years (rounding to 22). "One skilled in the art" (a common refrain in patent documents) can extend the maximum range in a similar manner; in this case, it would be necessary to further extend the age range to obtain the desired span of 10 years. The patent leaves

*Durand, Low, and Stoller, US Patent 6,272,467, p. 3.

unanswered the question of the "rational choice" of match ages for a 45-year-old client.

At this point, the database of members is searched for potential matches. First, many profiles are eliminated based on a major incompatibility: for example, age, smoking status, or not being of the preferred gender. An important feature is a *two-way compatibility check*: a smoker might be indifferent to whether or not their partner smokes, but any potential match who indicates a preference for nonsmokers would be eliminated at this point. Likewise, a male with a preference for female partners would be eliminated from a list prepared for a homosexual male. The remaining members are then filtered according to the stated preferences of the client, and a list of matches is produced.

Compatibility Scores

If online dating sites only provided lists of matches, few would pay for the service, and only free services would be successful. What makes people spend more than $2 billion a year on such sites is that potential partners are ranked according to some algorithm to produce a *compatibility score*, and only those profiles exceeding a certain compatibility threshold are presented to the client. The Spark Networks Services patent is unusually forthcoming in the details and is arguably a good model for how patents based on mathematical algorithms should be structured: the details of the algorithm are explicitly laid out, which make questions of infringement easy to answer. At the same time, the exact scope of the patent is clear, which allows for fundamental improvements to also be eligible for patents.

First, the address of a member is compared with that of the client, and the difficulty of meeting is assigned a value from 0 (very easy) to 60 (very difficult); a value over 20 indicates that it would be challenging for the client and match to meet, and these members would be eliminated from consideration as matches. This geographic information is converted into a point value using the formula

$$300 \times (1 - \text{Difficulty} \times 0.04)$$

Thus, a member deemed to have a difficulty of 15 (indicating somewhat inconvenient to meet) would receive $300 (1 - 15 \times 0.04) = 120$ points; a member having a difficulty of 20 would receive the minimum of 300

(1 – 20 × 0.04) = 60 points. In contrast, a member with a difficulty of 5 (indicating they were very convenient to meet) would have a difficulty of 300 × (1 – 5 × 0.04) = 240 points.

Next, 300 points are assigned if the member's age is within the range *originally* specified by the client. A 27-year-old match would receive 300 points for being within the 45-year-old's original "26 to 28 years old" range. Since this range was deemed too narrow, it was extended by 5 years downward (and some unstated amount upward); a match that fell into the extended range would receive fewer points based on the formula

$$300 \times (1 - \text{Fraction of Extended Range} \times 0.2)$$

based on the fraction of the extended range required to include the match. In this case, the range was extended by 5 years, so a 24-year-old (2 years below the originally requested minimum of 26) requires 2/5 = 0.40 of the extended range, and her compatibility score would be 300 × (1 – 0.4 × 0.2) = 276 points. This formula gives between 240 and 300 points based on age preferences.

Finally, between 35 and 400 points are assigned on the basis of physical build. First, a member's height and weight are converted into a build class, based on actuarial tables for obesity and assigned a value of 1 (small) through 5 (XX-full). Next, the client's desired build is adjusted upward by 1 (if the person is deemed X-full) or 2 (if the person is deemed XX-full). If a potential match has a build equal to or lower than the user's adjusted desired build, a *build factor* of 1.0 is assigned to the match. Otherwise, a difference of 1, 2, 3, 4, or 5 build classes is assigned a build factor of 0.90, 0.26, –0.57, –2.67, respectively. A similar build factor is computed for the potential match. Then, the build component of the compatibility score is computed using the formula

$$400 \times (0.7 \times \text{Client Build Factor} + 0.3 \times \text{Match Build Factor})$$

For example, suppose our client, an overweight, middle-aged man, has a build class of 4 (X-full) with a preference of 1 (small). First, his desired build would be adjusted upward to 2 (medium).

Now consider a potential match: a woman with a build class of 1 (small) who seeks a man with a build class of 1. Since her build class is small, her desired build is not adjusted and remains at 1. Since her build class is less than the client's, the client build factor is 1. But since *his* build class is 4, while her desired build class is 1, there is a 3 build class difference, corresponding to a match build factor of 0.26. As a potential match, her compatibility score will be 400 × (0.7 × 1 + 0.3 × (−0.57)) = 211.6, which would be rounded to 212.

The geographic, age, and build scores are added together to form a basic compatibility score, which will be between 335 and 1,000 points, with higher values indicating greater compatibility. This score is then adjusted based on the presence or absence of other factors.

What is most surprising about these adjustments is that they are very small in comparison to the points awarded for geographic location, age, and build. Leaving aside bonuses for recent activity, the most important secondary factors revolve around smoking and children. If the client is a nonsmoker who indicates "no preference" whether a match is a smoker or not, then a match who is a nonsmoker would receive 14 bonus points. Similarly, if the client prefers "no children," then a match with no children would also receive 14 bonus points. Meanwhile, if client and match indicate they like the same types of activities (for example, dinner and a movie vs. a sporting event), they receive just 6 bonus points. In contrast, a male match is awarded 30 points per inch of height (maximum of 60 points) if they are taller than a female client, and penalized *500 points* if they are 3 or more inches shorter (or 30 points per inch if they are from 1 to 3 inches shorter). All other things being equal, a woman seeking a male nonsmoker without children would find a male smoker with children ranked higher than a male nonsmoker without children — as long as he was two inches taller than the client.

What Men and Women *Really* Want

The fact that 700 out of 1,000 points on the compatibility scale are determined by physical characteristics like age, weight, and height, should cause us dismay: Are we really so shallow as to make judgments based on physical appearances? Or do dating services *assume* that physical appearance is important and, by doing so, *make* it relevant?

An early study on what men and women want was conducted in 1939, by Reuben Hill of Iowa State College. Hill surveyed 600 students in an attempt to find what men and women looked for in a potential spouse. Participants ranked 18 characteristics on a scale from 0 (unimportant) to 3 (indispensable). Both sexes declared a strong preference toward a person with dependable character, emotional stability, and pleasant disposition: these "mature choices on which companionship would normally be based"* all ranked among the four most indispensable traits by both men and women. Meanwhile, physical appearance and financial prospects ranked very low, with political affiliation ranked least important.

The most interesting results were the gender differences in the less important traits. For example, men considered "good looks" to be more important than "good financial prospects," while women reversed their ordering. Again, neither of these traits were ranked very highly (14th and 17th for men, 17th and 13th for women).

We might wonder whether a study conducted in prewar America is relevant to the internet generation. However, Hill's study has been repeated many times, and only minor changes have occurred in the past 60 years. A 2001 incarnation, by David Buss (the University of Texas at Austin), Todd Shackleford (Florida Atlantic University), Lee Kirkpatrick (the College of William and Mary), and Randy Larsen (Washington University) showed that character, stability, and pleasing disposition were still deemed the most important traits. Meanwhile, "good looks" and "financial prospects" continue to show a distinct gender difference, with men ranking them at 8th and 13th, and women ranking them at 13th and 11th.[†] We might take solace in the fact that "good looks" and "financial prospects" are very far down on the list of things we consider important in a match: perhaps we're not as shallow as it seems.

Or are we? In 2008, Paul Eastwick and Eli Finkel of Northwestern University studied the behavior of men and women in *speed dating* situations: arranged social gatherings, where individual men and women meet for a brief period of time (typically under 5 minutes) and then switch to a new partner. In that time interval, they form an opinion of their opposite, and indicate whether they wish to pursue further contact (yes) or not (no). Sometimes a third possibility, a *missed match*, is allowed: this is a no, with the possibility of being changed to a yes. We'll

*Reuben, "Campus Values in Mate Selection," 557.

[†] The items that lost the most ground were chastity and sociability. Interestingly, religious background and social status remained virtually unchanged. Political affiliation remained least important.

indicate this as no/yes. If both partners indicated yes, or if one partner answered yes and the other no/yes, the organizers would put them in contact with each other. Any other pairing (yes and a straight no, or even a no/yes with a no/yes) would result in no further contact.

Eastwick and Finkel recruited 163 students (81 men, 82 women) to participate in a speed dating event. The participants completed questionnaires beforehand, identifying qualities they sought in a match. These included factors such as physical appearance, earning prospects, and personability. Consistent with the Hill study and its successors, both sexes ranked personability as more important than appearance or earning prospects; also consistent with the Hill study and its successor, men ranked appearance more important than earning prospects, while women ranked earning prospects more important than appearance.

However, when Eastwick and Finkel analyzed results afterward, they found that physical appearance and earning prospects more reliably predicted matches than personality. Moreover, the gender difference disappeared. For *both* men and women, physical appearance proved the most important factor, and for *both* men and women, earning prospects proved second most important. No matter what people say, both men and women want the same thing: a good-looking partner who can make a lot of money.

One limitation of Eastwick and Finkel's study is the participants rated the attractiveness of the others, so it's possible that a participant rated another as attractive because they were interested in pursuing a relationship and not the other way around. Thus, a more useful study would rate the attractiveness of the participants by an impartial third party. This was done in 2014, by Andrea Meltzer (Southern Methodist University), James K. McNulty (Florida State University), Grace Jackson (UCLA), and Benjamin Karney (UCLA), who found that marital satisfaction, particularly among men, is dependent on the physical attractiveness of the spouse, as judged by uninvolved third parties (graduate students).

For better or for worse, appearances do matter, and however uncomfortable we might be with this observation, ignoring this fact because it is socially unpalatable is equivalent to not building a tornado shelter in Oklahoma because you object to inclement weather. A service trying to find a suitable match for a client must account for physical appearance and financial prospects.

The Eye of the Beholder

Again, financial prospects are relatively easy to quantify. But it might seem that standards of beauty vary too much to allow for a meaningful comparison or, at the very least, limit such comparisons to questions about height, weight, and age. However, there are evolutionary reasons to believe that some universal standards exist, particularly those that correlate to likely fertility. The problem is determining what those standards are.

In 2006, Hatice Gunes and Massimo Piccardi, at the University of Technology in Sydney, Australia, conducted a study of facial beauty (Meltzer's group also ranked spouses by facial beauty). They began with 215 female faces of different ethnicities and had them ranked by 48 human judges, drawn mostly from Europe (29) and Southeast Asia (8), though some also came from India (2), the United States (4), Latin America (2), Australia (2), and Africa (1). The judges were of varying ages and genders and ranked the faces on a scale of 1 through 10.

As expected, there was considerable variability in the assessment of facial beauty. However, the judgment for each face had a strong *central tendency*, which is to say that many of the judges gave the same or similar rank to the face. The agreement among the judges suggests that there is some objective way to measure beauty.

But what were the judges focusing on? For centuries, philosophers, mathematicians, and artists have claimed beauty is based on *mathematical proportions*. The most infamous of these is known as the *golden ratio*, which arises (among other places) in connection with regular pentagons: it is the ratio of the length of a diagonal of a regular pentagon to its side and is equal to $\frac{1+\sqrt{5}}{2}$, which is approximately 1.618. Most of the so-called evidence for the golden ratio's aesthetic superiority comes from finding *some* pair of measurements in a work of art that leads to the golden ratio. However, these claims are dubious for the following reasons.

Consider the human face. There are a large number of landmark points on the human face: the top of the face, the tip of the chin, the nostrils, the tip of the nose, the pupils, the eyebrows, the lips. There are others, but suppose we focus on just these seven points. There are 21 measurements we could take: for example, tip of the chin to the eyebrows, lips to nostrils, or nose to the top of the face. From these

21 measurements, we can produce 210 distinct ratios. Given the relatively narrow possibilities for what these ratios could be, it's not too surprising that some of these fall into near approximations to the golden ratio.

While we can't rely on a simple-minded decision of whether certain facial measurements fit the golden ratio, we might be able to conclude something about facial measurements in general. To that end, we can employ *data mining*.

As its name suggests, data mining tries to find valuable nuggets of information from a large database. In real-world mining, we begin with a large volume of ore and progressively refine it, removing the dross to increase the concentration of the valuable materials. We do the same thing in data mining by creating a *decision tree*: a sequence of yes/no questions that splits our data into smaller and smaller sets with progressively higher concentrations of the desired object.

We can illustrate creating a decision tree as follows. Suppose we represent our data as a collection of black and white beans, with the black beans corresponding to the items of interest. Our goal isn't to find *all* the black beans; rather, we want to find a large set of them. We could go through the beans, one by one, and select a large enough set. However, if the fraction of black beans is small, we'd have to go through quite a few beans before finding enough to fill our set (Figure 4.1).

Instead, we might proceed as follows. First, dump all the beans onto a table. Then find a line separating the beans into two piles whose compositions are as different as possible. Finding such a line requires some trial and error. For example, we might draw a vertical line through the middle (Figure 4.1, Poor Division). On the left side, 40% (8 out of 20) beans are black, while on the right, 30% (6 out of 20) are black. While there is a difference in composition, we might be able to do better, so we try a horizontal line through the middle (Figure 4.1, Good Division). This time, 10% of the beans above the line are black, while below the line, 60% are black. Since this division yields a greater disparity in composition, we'll make "Are the beans below the vertical line?" the first branch (or *decision node*) in our tree. Note that we've phrased this as a yes/no question, instead of the more natural "Are the beans above or below the vertical line?" Moreover, for convenience, we'll word our questions so that yes leads us to the set with the greater proportion of what we seek.

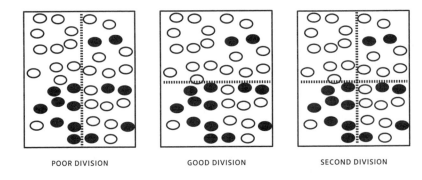

POOR DIVISION   GOOD DIVISION   SECOND DIVISION

Figure 4.1. Mining Black Beans

One of the things that makes constructing decision trees challenging is that the *order* in which we make our decisions matters. Splitting the pile vertically *first* didn't produce two sets with markedly different compositions. But if we split the pile vertically *second*, we will. If we take the beans below the horizontal line (which are "refined" so that 60% are black) and split this set vertically (Figure 4.1, Second Division), we find that the set on the left will be further refined so that about 89% of the beans are black. We might make "Are the beans to the left of the vertical line?" the next node in our tree; again, we've phrased this as a yes/no question, where yes leads us to a greater concentration of the items of interest. In this way, we can ask a sequence of yes/no questions that lead us to a pile with a high concentration of black beans.

How can we apply this problem to measuring facial beauty? Gunes and Piccardi focus on 16 measurements:

A. Top of face to tip of chin
B. Pupils to tip of chin
C. Top of face to nostrils
D. Top of face to pupils
E. Nostrils and tip of chin
F. Pupils to central lip line (between lips)
G. Central lip line to tip of chin
H. Pupils to nostrils
I. Nostrils to central lip line
J. Top of face to eyebrows

K. Eyebrows to tip of nose

L. Tip of nose to tip of chin

M. Tip of nose to central lip line

N. Central lip line to tip of chin

O. Horizontal pupil-to-pupil distance

P. Width of face

These 16 measurements can be used to find 120 different ratios. Gunes and Piccardi chose 14 of these ratios, deemed important by supporters of various aesthetic theories and, more pragmatically, by plastic surgeons. Next, they randomly separated the 215 faces into two sets: 165 images became part of the *training set*, with the remaining 50 forming the *test set*.

Now take any of the 14 ratios: for example, the ratio of J (the distance between the top of the face to the eyebrows) to A (the length of the face). Pick a value for this ratio: for example, 1 to 3. Then separate the training set into two parts: those where the ratio is more than 1 to 3 and those where the ratio is less than or equal to 1 to 3. If the two parts have a substantial difference in the fraction of high-ranked ("beautiful") faces, then we can form a decision node using the question, "Is the ratio of J to A more than 1 to 3?" If the two parts don't exhibit a substantial difference, then change the value of the ratio. If there's no value that yields a substantial difference, choose a different ratio.

Again, this first node splits the data into two parts, one of which has a higher concentration of "beautiful" faces. We can then try further refinement by choosing a different ratio and value: *If* the ratio of J to A is more than 1 to 3, our next decision node might be to determine whether the ratio of H to I is greater than 3 to 2. The 14 ratios focused on by Gunes and Piccardi allow for 14 successive refinements of the data set. With some minor modifications, this will not only identify "beautiful" faces but also return an actual rating on the 1 to 10 scale.

You might suspect that the decision tree is rigged: after all, the actual ratings are known, so what's to keep the researchers from setting the decision tree to reproduce these values? This is the purpose of the 50 values of the test set: we can run these 50 values through the decision tree and have it return a ranking; if the rankings agree (or, at least, disagree no more than the human judges themselves do), then the decision tree is validated. Gunes and Piccardi's work suggests that this approach to measuring facial beauty will produce meaningful results.

Every Accounting for Tastes

Still, we may hesitate to use this approach to evaluate online dating profiles. After all, just because someone else judges a person to be an "8" doesn't mean that *you* will make the same evaluation.

It would be easy to modify the approach of Gunes and Piccardi to reflect personal preferences. The client might review photographs of several hundred individuals (not necessarily members) and rank their physical appeal on a scale of 1 to 10, and then data mining strategies could be used to construct a decision tree for the client's personal preferences. But as the saying goes, never make it hard for someone to give you money: requiring a member to view and rank the hundreds of photographs necessary to produce a reliable decision tree could take hours, and even with the best intentions, a client might not complete this task. And even if he did, there's no guarantee that the last photographs would be ranked with the same care as the first few.

US Patent 8,369,582, "Automatic Appeal Measurement System," granted February 5, 2013, to Eastman Kodak Company, and invented by Andrew Blose and Peter Stubler, two computer scientists, and Joseph Manico, a photographic scientist, suggests an alternate approach. Rather than having the member rank photographs of several hundred individuals, imagine having a group of *proxy observers* rank an even larger set of thousands of photographs, not necessarily members. Using the previously discussed methods, a decision tree could be created for each of the proxy observers, and, as the decision tree is based on a large set of data, it would likely be a reliable predictor of ranking.

To account for differences in taste, the client is asked to judge the physical appeal of a *small* set of people. Based on the responses, the proxy observers are assigned a client-specific set of *weights*, with greater weight assigned to proxy observers whose judgments are more similar to that of the client. To predict a client's ranking of a member, that member's photograph is ranked by the decision trees of the proxy observers, and these ranks are multiplied by the weights and summed.

How can we determine the weights? We'll illustrate a standard approach in a system with just two proxy observers, to be assigned weights a and b. The client would be presented with photographs and asked to rank the physical appeal of the subjects; the photographs would also be run through the decision trees for the proxy observers. Each subject would receive three rankings: one from the client, and one from each of the proxy observers.

Suppose these ranks are 3, 4, and 3 (so the client, first proxy observer, and second proxy observer ranked the physical appeal of the subject a 3, 4, and 3, respectively). If the weighted ranks of the proxy observer accurately predict the client's rank, we should have

$$3 = 4a + 3b$$

There are an infinite number of possible values of a, b that will satisfy this equation. However, this is based on just one ranking. Consider a second subject, who receives ranks 8, 5, and 1. Since the weights don't change, our values of a, b should also satisfy the equation

$$8 = 5a + 1b$$

From algebra, we know that if we have two unknown values (the weights a, b) and two equations ($3 = 4a + 3b$, $8 = 5a + 1b$), we can find a unique solution to the weights.

However, only politicians and pundits base conclusions on two pieces of evidence. If facts are important, more is better, and so we might present a third subject, and obtain ranks 5, 2, and 8. This gives a third equation:

$$5 = 2a + 8b$$

Now we have *three* equations but only *two* unknowns. Again from algebra, we know that it may be *impossible* to find values a, b that satisfy all three equations.

So what can we do? Remember our goal is to produce a formula that will predict the client's ranking, based on the proxy observer's rankings. Consider any value of a, b: for example, $a = 0.5$, $b = 0.5$. The first subject received rank 4 and 3 by the two proxy observers; the values $a = 0.5$ and $b = 0.5$ would predict the client would assign rank 4 (0.5) + 3 (0.5) = 3.5. Since the client actually assigned rank 3, the *error* (in the sense of a difference between prediction and reality) is 3.5 − 3 = 0.5. For somewhat technical reasons, we use the *squared error*: $(0.5)^2 = 0.25$.

What about the second subject? Using $a = 0.5$ and $b = 0.5$, the formula predicts a rank of 5 $(0.5) + 1 (0.5) = 3$, but since the client actually assigned rank 8, the error is 3 − 8 = −5, with squared error $(-5)^2 = 25$. Finally, the formula would predict a rank of 2 $(0.5) + 8 (0.5) = 5$ for the third subject, which equals the client's rank, giving an error of 0 and a squared error of $0^2 = 0$. We've now made three predictions and obtained three squared errors. The *sum of the squared errors* is then 0.25 + 25 + 0 = 25.25. Since we'd like the error to be as small as possible, we want to find the values of a, b that minimize the sum of the squared errors.

The method of finding these values is known as *least squares regression* and was developed nearly simultaneously by three different mathematicians in the opening decades of the nineteenth century: German mathematician Carl Friedrich Gauss (1777–1855), French mathematician Adrien-Marie Legendre (1752–1833), and Irish-American mathematician Robert Adrain (1775–1843). Since then, it has become one of the most important tools in mathematical statistics. Applying this method, we find that the sum of the squared errors will be minimized when we let $a = 1.21$ and $b = 0.23$.

Now consider a subject whose photograph has not yet been presented to the client. Suppose we run a picture of the subject through the decision tree for the two proxy judges and find the first proxy observer would have ranked the member a 7, while the second would have ranked the member a 9. Based on these ranks and the assigned weights, we predict the client would rank this member 5 $(1.21) + 3 (0.23) = 6.74$.

Trolls

Blose and his co-inventors note that "the invention can be exercised to estimate the appeal of any type of 'individual,' subject to the availability of a valid similarity measurement that accurately estimates the similarity between individuals."* Because the scope of a patent is limited to what is included in the patent, it's common practice to include additional claims about what the patented device can be used for.

*Blose, Stubler, and Manico, US Patent, 8,369,582, p. 12.

The problem is that mathematical algorithms are extremely flexible and can be used in such a broad range of contexts that claiming an invention could be used to estimate the appeal of any "individual" based on the existence of a similarity measure could potentially give the holder of the Kodak patent a stranglehold on the use of similarity measurements. In theory, the holder of the Kodak patent could claim that a website that recommended movies based on your watchlist, a meal delivery service that created a menu based on your answers to food preference questions, or a matchmaking service that found romantic partners based on the results of a survey could all be infringements on their patent, because all of these measure the "appeal" of an "individual" based on a similarity measurement.

The habit of issuing overly broad patents is responsible for the existence of so-called patent trolls: companies that obtain very broad patents for the sole purpose of extorting licensing fees. For a time, the Kodak patent was held by Intellectual Ventures, "the most hated company in tech."*

Part of the success of patent trolls rested on the ability of a company to sue for patent infringement in virtually any court. Many found a friendly face in the court of Judge Rodney Gilstrap of the US District Court for the Eastern District of Texas, who saw nearly a quarter of all US patent cases. The power of the patent troll was diminished (but not eliminated) on May 22, 2017, when the US Supreme Court held that patent infringement suits must be tried where the defendants are incorporated, which makes it more likely that knowledgeable testimony can be obtained from members of the local community.

However, as long as overbroad patents are issued, patent trolls will exist, and the mere threat of a lawsuit is, in many cases, enough to force a settlement. Indeed, the costs of an actual trial are great enough that, more often than not, such infringement cases are settled out of court, leaving the question of what a patent actually covers unresolved.

*Jim Kerstetter, Josh Lowensohn, "Inside Intellectual Ventures, the Most Hated Company in Tech," CNET, https://www.cnet.com/news/inside-intellectual-ventures-the-most-hated-company-in-tech/. August 21, 2012. Intellectual Ventures subsequently sold the patent to Mountain Peak Ventures, a wholly owned subsidiary of Dominion Harbor Group, another company that exists primarily to hold patents.

Hamming It Up

One way to resolve the conflict between the endless generalizability of mathematics and the desire to limit the scope of a patent claim is to rely on a third criterion for patentability: usefulness. This is generally construed to mean that the invention has a use, however impractical; thus, we have wearable hamster trails ("pet display clothing"; US Patent 5,901,666), plastic sticks ("animal toy"; US Patent 6,360,693), and drink shades ("beerbrella"; US Patent 6,637,447).

However bizarre the invention, it must actually work; merely claiming that it will work is not enough. It's easiest to understand why in the context of pharmaceutical patents: a pharmaceutical company might produce a drug with no known uses. Without the usefulness requirement, they could receive a patent on the drug. On the one hand, they know of no uses for it. On the other hand, the existence of the patent would discourage others from finding a use for the drug, because any and all profits made from the drug would belong to the pharmaceutical company. Thus, patents must claim that the invention performs a specific task, and that claim must be supported.

The inventors of the Kodak patent suggest that proxy observers could be used to compare automobiles using the weighted Hamming distance. However, they provided no evidence that this approach would work. If we require that such claims of usability be supported, then the inventors would have to show that the method *could* be used to compare two automobiles, say, by compiling a database of automobiles and then running trials to show that proxy observers could be used to select automobiles that matched a user's preferences.

The *Hamming distance* was introduced in 1950 by information theory pioneer Richard Hamming (1915–1998): Given two vectors, the Hamming distance is the number of components that differ. The vector [1, 1, 0, 1, 2] and the vector [1, 1, 1, 3, 5] differ in three places, so they have Hamming distance 3.

Suppose each member of a dating site completes a five-item questionnaire with yes/no answers:

1. Do you smoke?
2. Do you like pets?

3. Do you like children?
4. Do you enjoy outdoor activities?
5. Do you consider yourself artistic?

Each member would then be completely characterized by the sequence of yes/no answers. Bob, a nonsmoker who likes pets but not children, likes outdoor activities and considers himself artistic, would answer, no, yes, no, yes, yes. Meanwhile, Alice, an artistic nonsmoker who likes pets and children but not outdoor activities, would answer, no, yes, yes, no, yes. Finally, Carol, an artistic smoker who doesn't care for pets or children but enjoys outdoor activities, would answer, yes, no, no, yes, yes.

This sequence of yes/no answers allows us to identify each person with a vector. Thus Bob could be associated with the vector [No, Yes, No, Yes, Yes], while Alice would have the vector [No, Yes, Yes, No, Yes], and Carol would have vector [Yes, No, No, Yes, Yes]. We can then use the Hamming distance to gauge compatibility. Two persons with identical responses would have Hamming distance 0, while two persons with completely opposite responses would have Hamming distance equal to the number of questions (in this case 5).

In this case, Bob and Alice answered questions three and four differently, so the Hamming distance between them would be 2. Similarly, Bob and Carol answered questions one and two differently, so the Hamming distance between them would also be 2.

There are some problems with using the simple Hamming distance. The biggest is that it treats all differences equally. As far as the Hamming distance is concerned, Bob and Alice are as compatible as Bob and Carol. However, Bob and Alice differ on the question of children and outdoor activities, while Bob and Carol differ on the question of smoking and pets. If Bob is amenable to changing his mind about smoking, but less likely to change his mind about children, then Bob would be more compatible with Carol than with Alice.

To reflect this, we can use a *weighted Hamming distance*. This can be accomplished by asking clients to rate the importance of characteristics on a *Likert scale*, which records the intensity of feeling. For example, the client might respond with a value from 0 (*unimportant*) to 5 (*extremely important*). While we could ask the client to assign these values to the different characteristics, we might again rely on proxy observers.

Suppose there are two proxy observers. Observer 1 assigns the weights 3, 2, 3, 1, 1 to the five characteristics. Remember, if two people have the same trait, the Hamming distance would be 0, while if they differ, the Hamming distance would be 1. The weights multiply these values, and the greater the weight, the farther apart the two profiles. The high values assigned to the first, second, and third traits indicate that for this observer, similarity of views on smoking, pets, and children are important. Note that this doesn't mean a match must have a particular view; rather, they have to have the same view. The second observer might assign the weights 0, 1, 1, 2, 6: similar views on smoking aren't important, but outdoor activities are, and artistic inclinations are very important.

Now consider Bob, an artistic nonsmoker who likes pets and outdoor activities but not children. To solve the problem of least squares in a meaningful way, Bob must rank more profiles than there are proxy observers. If we want to use two proxy observers to predict Bob's ranking of a client, we'd need to have Bob rank at least three profiles.

We might present Bob with three profiles and ask him to rank "how similar" they are to him. Since we want more similar persons to have lower distances we might ask Bob to rate profiles on a scale where 0 is the most similar and 10 is the least similar.* With this in mind, we might find the following:

1. Alice, an artistic nonsmoker who likes pets and children but not outdoor activities: Bob assigns her a 3 (similar).
2. Carol, an artistic smoker who doesn't care for pets or children but enjoys outdoor activities: Bob declares her to be very dissimilar (8).
3. Dana, a nonartistic nonsmoker who likes pets but not children or outdoor activities: Bob gives her a 5, indicating some similarity.

What about the proxy observer? Bob's profile is [No, Yes, No, Yes, Yes], while Alice's is [No, Yes, Yes, No, Yes]. She differs from Bob on traits 3 and 4 and shares the same views on the others. The first proxy observer, who assigned weights 3, 2, 3, 1, 1, would give Alice a distance of $3 + 1 = 4$

---

*In practice, we might give Bob a scale of choices, from "Completely Dissimilar" to "Completely Similar," and convert a response into a number from 0 to 10.

(the sum of the weights of the two divergent traits); the second proxy observer, who assigned the weights 0, 1, 1, 2, 6, would give Alice a distance of $1 + 2 = 3$. By a similar determination, we find that Carol would be scored $3 + 2 = 5$ by the first observer and $0 + 1 = 1$ by the second; Dana would be scored 2 and 8 by the two observers.

If we apply the least squares method, we find we should assign the first proxy observer a weight of 1.21, while the second proxy observer should be assigned a weight of 0.23. In effect, this means that the first proxy observer's judgments are more in line with Bob's own. This allows us to predict Bob's similarity to Ellen, an artistic nonsmoker who likes children and outdoor activities but not pets: she differs from Bob on characteristics 2 and 3. The first proxy observer would have assigned her Hamming distance $2 + 3 = 5$, while the second would have assigned her Hamming distance $1 + 1 = 2$, and we would predict Bob would rank her similarity as $5\left(1.21\right) + 2\left(0.23\right) = 6.51$.

Sevens Date Sevens

While the preceding method would allow us to judge how similar two people are to each other, is this enough for successful matchmaking? If we go by pithy aphorisms, it isn't. Familiarity breeds contempt, while opposites attract. But do these bits of folk wisdom actually work? Should we seek partners who are our opposites and set up dates between PhD researchers and high school dropouts, recommend extreme hikers to online gamers, and have capitalists marry communists?

An early study on preferences occurred in 1970, when Donn Byrne and John Lamberth, both of Purdue University, and Charles Ervin, of the University of Texas, asked 420 psychology students at the University of Texas to complete a 50-question survey designed to elicit information about their personalities. The researchers then found the male-female pairs with the *highest* number of similar answers and the pairs with the *lowest* number of similar answers. Some additional criteria were imposed: the male of the pair had to be taller than the female, neither person could be married, and, as the study was conducted in Texas in 1970, interracial pairings were rejected.

In all, 24 "high similar" pairs gave the same responses on more than 67% of the questions, and 20 "low similar" pairs gave similar responses on less than 40% of the questions. Each pair was given 50 cents (worth roughly $3 today) and sent to the student union to go on a 30-minute "Coke date," and when they returned, the researchers asked them a number of questions. Rather than ask the individuals to report on their level of attraction for each other, which would raise the perennial problem of objectively reporting one's own sentiments, the researchers instead noted where the members of the pair sat during the interview: the closer they sat, the higher the level of attraction. The subjects were then separated and asked to rate the other person's intelligence, physical attractiveness, and desirability as a date or a spouse.

The results showed that similarity had a profound effect: not only did similarity lead to greater attraction (as determined by the distance between the members of the pair in the after-interview), it was also correlated to a higher judgment of the partner's intelligence and desirability as a date and a spouse. When it comes to seeking out dates or mates, opposites *don't* attract.

However, as with the Hamming distance, we shouldn't assume that all similarities are treated equally: it may be critical that your partner shares your religion but less important that they share your political views. One way to measure the importance of similarity or dissimilarity is to find the *bounding strength* of a characteristic: how much more likely two persons who share the characteristic will be to contact each other.

For example, about 2% of the world's population is redheaded. Suppose we consider a group of 10,000 heterosexual couples.* Since 2% of the men are redheaded, the group would contain about 200 redheaded men. These 200 men would have 200 female partners, and since 2% of the women are redheaded, we'd expect this second group to include about four redheaded women. Thus, we'd expect to see four couples consisting of a redheaded man with a redheaded woman.

Suppose redheaded men and women seek each other out. Then we'd see more than four couples consisting of two redheads. We can use the ratio between the observed number and the expected number as

*The heterosexuality of the couple plays no role in the following: mathematics doesn't care about gender, and we could talk about "the first member of the couple" and "the second member of the couple." To avoid this wordier construction, we'll assume that the couples are heterosexual, so we can talk about the man and the woman in a couple.

a measure of this tendency. Conversely, if redheaded men and women avoided one another, we'd see fewer than four couples in which both are redheaded.

For example, if we observed 8 couples consisting of 2 redheads, we'd observe twice as many as we expected to see. We'd say that *bounding strength* of red hair is 2. Or if we saw only 1 pair of redheads, then we saw $\frac{1}{4}$ as many couples as we expected to see, giving us a bounding strength of $\frac{1}{4}$ = 0.25.

A bounding strength greater than 1 is evidence for *homophily* (seeking sameness), while a bounding strength of less than 1 can be taken as evidence for *heterophily* (seeking difference). "Opposites attract" is a claim that heterophily is the norm, while "sevens date sevens," a theory that one should seek a partner with a similar level of physical attractiveness, is an expression of homophily.

In 2005, Andrew Fiore and Judith Donath of MIT's Media Lab studied the 65,000 users of an online dating system as they interacted between June 2002 and February 2003. They measured two quantities: first, the fraction of contacts made to someone with a shared characteristic (Bob sends a note to Alice, hoping to meet); and second, the fraction of reciprocated contacts between persons with shared characteristics (Alice sends a note back to Bob, agreeing).

Their work provides strong evidence of homophily. The highest bounding strengths centered around marital status and a desire to have children. We'd expect to see 31.6% of reciprocated contacts to be between users of similar marital status (single, divorced, widowed) and 25.1% to be between users with similar desires to have children. Instead, 51.7% of the two-way interactions occurred between users of similar marital status, and 38.7% occurred between users with similar desires to have children, corresponding to bounding strengths of 1.64 and 1.54, respectively.*

As expected, physical factors also play a significant role. We'd expect 19.2% of contacts to occur between users of the same self-described physical build (athletic, petite, bodybuilder, average), but, in fact, this occurred 25% of the time, corresponding to a bounding

---

*The percentage of one-way contacts, where one user did not respond to an overture, was even higher.

strength of 1.28. Finally, Fiore and Donath found that, while we'd expect to see 37.6% of reciprocated contacts between those with similar (self-designated) attractiveness, the actual percentage is 46.1%, giving a bounding strength of 1.23 and providing evidence for the "sevens date sevens" hypothesis.

Sex in the City

Fiore and Donath's work suggests that homophily is a useful way to measure compatibility. Since we can represent a person's characteristics in vector form, it follows that any method of comparing two vectors could be turned into a useful tool to measure compatibility.

We've already introduced two ways to measure the similarity between two vectors: the dot product and the Hamming distance. But there are many others. One of the more common is the *Manhattan metric*, also known as the *taxicab distance*. Imagine you're in a city whose streets form a perfect grid, and you want to find the distance between two points. The Euclidean distance between two points is measured along the straight line path connecting the two points. However, unless there's nothing but open ground between the two points, the straight line path will generally be impossible to follow: you'd have to contend with buildings and trees. Thus the Euclidean distance will often be irrelevant.

Instead, we would use the Manhattan metric: the sum of the differences. You're at 5th Avenue and 42nd Street, and a friend is at 8th Avenue and 46th Street, the difference between you will be 3 avenues and 4 streets, giving a Manhattan distance of 3 + 4 = 7. As with the Hamming distance, we might weight the components. If going from one street to the next required dealing with traffic, we might count the street-to-street distance as twice its actual value. This would give us a weighted Manhattan distance of 3 + 2 × 4 = 11.

On August 10, 2000, J. Galen Buckwalter, Steven R. Carter, Gregory T. Forgatch, Thomas D. Parsons, and Neil Clark Warren, a group of psychologists, submitted a patent application for a system that uses weighted Manhattan distance to determine compatibility. The same month, they launched eHarmony. On May 11, 2004, they received US Patent 6,735,568, "Method and System for Identifying People Who Are

Likely to Have a Successful Relationship." A few months later, they se-
cured an infusion of more than $100 million in venture capital money.

eHarmony touts a "scientific algorithm" for finding soul mates. Since
the fortunes of the company (at $60 per member per month) depend
on providing a service different from its competitors, the exact details
of these scientific algorithms are closely guarded secrets. However, the
patent application suggests the general outline of the algorithm.

First, users complete a questionnaire, which allows eHarmony to
compute an *individual satisfaction estimator* (ISE). This is computed
using a weighted sum of personality factors. These aren't disclosed in
the patent, but presumably relate to the "29 dimensions of compatibility" touted in eHarmony's advertising literature. This allows potential
clients to be classified into categories, such as "unlikely men" or "good
women," based on the likelihood of forming a satisfactory relationship
with another person. "The unlikely classification may indicate that
the user is unlikely to be happy in any relationship. . . . The matching
service may choose not to provide service to people who fall within particular classifications."* An unusual feature of eHarmony is that it will
reject a significant fraction of those who seek to use its services. While
the company has drawn criticism for this practice, the alternative is
to take someone's money even though, in the company's opinion, they
have little chance to benefit from the service. In this respect, eHarmony
is refreshingly ethical.

Next, the *couple satisfaction index* (CSI) is computed for the couple
consisting of the client and a match. This is found using a weighted
Manhattan distance, again based off a number of personality factors
(again, not disclosed by the patent, but presumably related to the "29
dimensions of compatibility"). If the CSI surpasses some threshold
value, the two members of the couple are recommended to each other.

Big Hands, Big Feet

Although eHarmony gathered headlines with its scientific algorithm
for matchmaking, it is a latecomer to the field. Match.com, launched
in 1995, is actually the most important player in the field. On June 5,

*Buckwalter et al., US Patent 6,735,568, p. 9.

2012, Match.com received US Patent 8,195,668 for "System and Method for Providing Enhanced Matching Based on Question Responses," developed by Ricky Drennan, Sharmistha Dubey, Amanda Ginsberg, Anna Roberts, Marty Smith, and Stanley E. Woodby Jr.

As with eHarmony's scientific algorithm, the details of Match.com's algorithm are shrouded in secrecy, but again, the patent allows us to glean some information. Members answer a series of questions that allow them to be classified according to one of four personality types: Explorer, Builder, Negotiator, and Director. We might use homophily to find suitable matches, but as the inventors note, homophily might not always make for a good match: "Todd may be extroverted and excessively social, but if his mate shared this character trait, this situation may be unworkable."*

Instead, they focus on *complementarity*. Instead of relying only on homophily, personality types are matched according to a set of relationship rules. Explorers are matched with other Explorers, but Builders are matched with Directors.

However, the best-laid plans are likely to leave us only grief and pain, if they are based on a false self-assessment: we must see ourselves as others see us. To avoid the problems inherent with self-evaluation, Match.com's patent uses a rather unusual way to measure personality: biometrics. At one point, it was believed that a person's character could be determined from their physical features. Vestiges of these medieval beliefs remain in our language and culture: *gauche* and *sinister* come from the French and Latin words for "left," and wedding bands are worn on the ring finger because of a belief that a nerve passed directly from that finger to the heart.

One might expect the age of reason — not to mention a better knowledge of anatomy — to eliminate these beliefs, but the quest to quantify led to the development of phrenology in the nineteenth century, which claimed to be able to predict personality through the shape of a person's skull. Better measurements and more rigorous studies have thoroughly debunked phrenology as a legitimate science, but even today, it's standard mythology among middle schoolers that you can determine a classmate's personality from the shape of his or her fingers.

*Drennan et al., US Patent 8,195,668, p. 8.

While this seems to be little more than an updated version of phrenology, there is some evidence for the claim. Helen Fisher, an anthropologist at Rutgers University, has investigated what, if any, links exist between physical traits and personality. Unlike phrenology or ring wearing, which are based on erroneous beliefs about human anatomy, Fisher's work focuses on the effects of sex hormones on physical development and has become an integral part of the algorithm for Chemistry .com (a premium service available from Match.com).

Testosterone has drawn the most attention from researchers. While both men and women produce testosterone, men produce far more. The effects of testosterone on behavior are well known to popular culture: action movies are sometimes derided as "testosterone fests," reflecting the belief that violence and aggression are particularly male traits. Studies suggest a number of other traits are related to elevated levels of testosterone. Among them are having greater focus, being less respectful, and possessing poor verbal fluency.*

If it is true that some of our personality is determined by testosterone levels, it follows that an accurate measure of testosterone levels can be used to assess character traits. The obvious way to do this is through a blood sample, and while some services (notably Genepartner.com) use DNA and immunotyping to identify potential matches, most choose less invasive procedures.

Fisher's work focuses on measuring the ratio between the second and fourth digits of the hands. For both men and women, the second digit (forefinger) is typically shorter than the fourth (ring finger). However, in men the ratio is smaller. This difference is generally attributed to fetal exposure to testosterone.

Can we use the digit ratio (referred to as 2D:4D) as a way to predict traits? Several recent studies suggest it's possible. In 2009, John M. Coates, Mark Gurnell, and Aldo Rustichini, of the University of Cambridge, found a strong correlation between the ratio and success as a high frequency stock trader (the lower the ratio, the better the stock trader). Then in 2012, Shin-ichi Hisasue and Shigeo Horie, of Teikyo University; Shoko Sasaki, of the University of Tokyo; and Taiji

---

*Or being nitpicky, assertive, and concise. Like everything else, personal behavior is interpreted differently by different people.

Tsukamoto of Sapporo Medical University, all in Japan, noted that the ratio correlated with female gender identity (the higher the ratio, the more a person self-identified as female), and in 2013, Jeremy Borniger, Adeel Chaudhry, and Michael Muehlenbein, of Indiana University in Bloomington, published a study that suggested that females with a lower 2D:4D ratio tend to have greater musical abilities.

None of these studies claim that having a low 2D:4D ratio *causes* you to become a stock trader or violin player. Rather, they identify correlations between the ratio and these characteristics: it's like determining the annual income of someone by looking at the car they drive. And while it's true that some people drive very expensive cars in spite of a low annual income, the fact remains that drivers of expensive cars tend to have higher annual incomes. In the same way, those with a low 2D:4D ratio tend to have certain personality traits. Since the ratio is a measurable quantity, it provides an objective way to characterize a member.

A bigger problem is that biometrics requires painstakingly accurate measurements. For example, in the Bloomington study, the 2D:4D ratio among female music students was 0.968 (indicating the forefinger was 96.8% of the length of the ring finger), compared to 0.971 for nonmusic students. Thus, if a woman's forefinger has a length of exactly 70 millimeters, we might measure her ring finger to have a length of 72.1 millimeters, giving her a 2D:4D ratio of 0.971, comparable to the nonmusic students. But if we mismeasure her ring finger at 72.3 millimeters, her 2D:4D ratio would be 0.968, comparable to the music students. The difference between the two measurements is only 0.2 millimeters — slightly more than the thickness of two sheets of paper. While such accuracy is possible under laboratory conditions, having people self-report will almost certainly lead to misclassification.

Does It Work?

The current generation of online dating sites claims to use scientific algorithms to find suitable matches for members. But this is tantamount to being able to measure love and happiness, and we might wonder if such a thing is even possible. Is there *any* way to predict whether a given couple will enter a satisfactory relationship?

We might ask, *Do these algorithms work?* In 2012, Eli Finkel, of Northwestern University; Paul Eastwick, of Texas A&M University; Benjamin Karney, of the University of California, Los Angeles; Harry Reis, of the University of Rochester; and Susan Sprecher, of Illinois State University attempted to answer this question.

There are two challenges to answering this question. First, the companies keep the details of the algorithms secret. In and of itself, this isn't a problem, since many problems in psychology and biology contend with such "black box devices" whose operational details are unknown. The problem arises if a company changes its algorithm: many such companies boast of their constant efforts to improve their services. This may serve customers better, but it also means that data collected before the change can't be meaningfully compared to data collected after the change.

A more fundamental problem is the following: eHarmony claims responsibility for more than 400 marriages a day. If their scientific algorithm is effective, these 400 marriages are somehow better than marriages not arranged by eHarmony. One obvious way to measure the quality of a marriage is to determine how long it lasts. But we can't just compare the average duration of an eHarmony marriage to the average duration of a non-eHarmony marriage. What we need to do is to compare the average duration of an eHarmony marriage to the average duration of a non-eHarmony marriage between people who are as similar to eHarmony users as possible: in effect, we need to find spouses who differ from eHarmony users *only* in that they are not eHarmony users.

But what is similar? We might look for people of similar age, occupation, geographic location, and so on. However, there is one crucial factor that distinguishes eHarmony users from non-eHarmony users: the willingness to pay $60 each month to find a match. This willingness to pay may reflect underlying personality traits: for example, it may be that eHarmony users are willing to go the extra mile to make a relationship work. If *this* is the factor that determines whether a marriage lasts, then eHarmony marriages will tend to last longer regardless of the efficiency of its algorithms.

The lack of hard data, as well as the overall difficulty of disentangling the various causes for what make marriages succeed or fail, led Finkel and his colleagues to conclude: "No compelling evidence supports matching sites' claims that mathematical algorithms work — that they foster romantic outcomes that are superior to those fostered by other means of pairing partners."*

Still, even if it turns out that matching services are no more effective than singing karaoke at the office Christmas party, this does not mean that they are a waste of money: they are, after all, a *service*. You could cook all your own food and wash your own clothes. Or you could eat at restaurants and drop your laundry off at a cleaner. You pay for the convenience of not having to cook or clean for yourself.

*Finkel et al., "Online Dating," 3.

5

〰〰〰〰〰〰〰〰

The Education Revolution

There must be an "industrial revolution" in education, in which educational
science and the ingenuity of educational technology combine to modernize
the grossly inefficient and clumsy procedures of conventional education.

SIDNEY PRESSEY
Psychology and the New Education

In 1589, Englishman William Lee of Calverton, in Nottinghamshire,
invented the stocking frame, a mechanical device that mimicked the actions of a human knitter and presented it to Queen Elizabeth I, hoping
to obtain a patent. Elizabeth demurred, fearing the device would cause
economic hardship among poor people, many of whom supplemented
their livelihood by hand knitting. Elizabeth's successor James I also declined to give Lee a patent, so Lee moved to France, where he gained
the patronage of Henry IV. Henry took the peculiar viewpoint that the
welfare of his citizens was more important than which religion they
happened to follow; as a result, he was assassinated by a fanatic in 1610.
With the loss of his patron, Lee died in Paris, in poverty.

Lee's brother, and the remaining machines, made their way back
to England. There, one of his apprentices, John Ashton, continued
to improve the machine. By the mid-1700s, a thriving mechanical
knitting industry had developed in Nottinghamshire. However, the
stocking frames themselves had grown too expensive for any individual

workman to own them: they were instead owned by wealthy men, who rented them out to workers; in turn, the workers produced textiles, which they sold back to the owners.

The system worked well — for those who owned the machines. But a recession, caused in part by the Napoleonic Wars, left frame owners reluctant to buy the finished product. At the same time, they held workers to the terms of the rental agreements, forcing many into bankruptcy. In 1811, bankrupted workers calling themselves Luddites (after a legendary founder, Ned Ludd who, like Robin Hood, hailed from Nottinghamshire) began destroying the stocking frames.

The government's response was swift and harsh: the British Army was deployed against the Luddites, and Parliament made deliberate destruction of the frames an offense punishable by death. By 1813, between 60 and 70 Luddites had been hanged, and the movement had been suppressed. Since then, the term *Luddite* has been used as a name for those who oppose technological progress because of the potential for economic dislocation.*

TORQing the Workforce

Most economists believe that technological advances create more jobs than they destroy. The problem is that the new jobs usually require different skills from the old jobs. The arc welder replaced by a robotic welder will take small consolation from the fact that the machine requires a technician to program and maintain it, because the arc welder will not, in general, have the skills to be the technician. So what is a newly unemployed welder to do?

The obvious strategy is to look for other welding jobs. However, this only delays the inevitable: once *some* welding jobs are done by machines, it's only a matter of time before *all* welding jobs are done by machines. A better approach is to look for jobs that match the skills, knowledge, abilities, and interests of the person, regardless of the actual job description. This information is available through O*NET, the Occupational Information Network, a database maintained by the US Department of Labor.

*In France, a less violent form of labor protest emerged: workers protested by "dragging their feet," or in the French idiom, "walking in wooden shoes [*sabot*]," which gave rise to the term *sabotage*.

Every job in the United States done for pay or for profit is assigned a Standard Occupational Code (SOC) by the Department of Labor. In some cases, these codes correspond to job titles: 53-2031.00 is "Flight Attendant." In others, the codes refer to specific responsibilities, independent of the actual title: 11-1011.00 is "Chief Executive," and applies whether the person's job title is "Aeronautics Commission Director" or "Store Manager" or "Liquor Stores and Agencies Supervisor." O*NET has detailed information on about 95% of the occupations available to workers in the United States.

Much of this information centers around the skills and knowledge necessary for the occupation. To determine this, career analysts and incumbents (those currently employed in the field) are surveyed to determine the importance of different knowledge domains, such as mathematics, design, or food production. These are ranked on a scale from 1 (not important) to 5 (extremely important). Next, if a knowledge domain is at least somewhat important (2 or more), the level of knowledge is then ranked on a scale of 1 (minimal knowledge) to 7 (expert knowledge). For food production, a level of 2 corresponds to the ability to keep an herb box in the kitchen, while a 6 corresponds to the ability to run a 100,000-acre farm. These raw scores are then converted into a value from 0 to 100. We might find that mathematics is very important to a cabinet maker (80), though the level is relatively modest (47). In contrast, a web administrator uses mathematics rarely (41), though at a more advanced level.*

We can perform a similar analysis for other job-related factors. Multilimb coordination (classified as an ability) is important to dancers (78) and even more important to pile drive operators (81), though dancers require it at a higher level (70, intermediate between operating a forklift and playing drums in a jazz band) than pile drivers (57).

The ability to characterize an occupation by a sequence of numbers allows us to describe jobs using vectors, where each component corresponds to an importance or level score for a skill, an ability, or

---

*Unsurprisingly, high-level mathematics (93) is very important (98) to postsecondary mathematics teachers. Somewhat surprisingly, mathematics is less important (95) to mathematicians, though they do require higher-level mathematics (97). And to postsecondary physics teachers, mathematics at an even higher level (99) is deemed more important (100, the maximum possible value).

a knowledge domain. This approach is the basis of TORQ, the Transferable Occupation Relationship Quotient, invented by Workforce Associates (now TORQworks) founder Richard W. Judy. Judy's storied career includes being an exchange student at the University of Moscow at the height of the Cold War (1958–1959), a stint as a professor of economics and computer science (University of Toronto, 1964–1976), and research fellow at the Hudson Institute (1986–2000).

At the Hudson Institute, he was director of the Center for Soviet and Eurasian Studies but later became director of the Center for Workforce Development. This piqued his interest in employee development and eventually led to his founding of TORQworks. On June 25, 2013, Judy received US Patent 8,473,320 for his method of analyzing jobs to assess the feasibility of changing from one career to another.

For illustrative purposes, suppose we focus on three professions (dancer, pile driver, and motorcycle mechanic) and three abilities (multilimb coordination, visual color discrimination, and hearing sensitivity). Each of the abilities has three scores: importance, level, and what Judy names the Combined Multiplicative Descriptor (CMD), obtained by multiplying the two. This means that each profession can be associated with three vectors:

- Dancer: Importance vector $[4.13, 2.25, 2.38]$; Level vector $[4.88, 1.88, 2.38]$; CMD $[20.15, 4.23, 5.66]$
- Pile driver operator: Importance vector $[4.25, 2.5, 2.62]$; Level vector $[4, 1.88, 2.38]$; CMD $[17, 4.7, 6.24]$
- Motorcycle mechanic: Importance vector $[3.25, 2.38, 3.88]$, Level vector $[3.12, 2.25, 3.88]$; CMD $[10.14, 5.36, 15.05]$

Now imagine a dancer wanted to change jobs. Which job is the most compatible with his or her current skills and abilities?

We're looking for compatibility or similarity, and we could use any of a number of vector similarity measures, such as those used to compare documents, images, or lifemates. Yet another possibility is to examine the *correlation* between the two vectors.

To understand correlation, imagine two seemingly unrelated quantities: for example, hemlines (the distance from the floor to the bottom edge of the fabric of a skirt) and stock market prices. Mathematicians would say both are *random variables*, meaning that the actual values are in practice unpredictable. Random does not mean reasonless: hemlines and stock prices rise and fall for very definite reasons. What makes these changes random is that they are *unpredictable*.

While the lengths and prices are random, it's conceivable that there is a correlation. *Positive correlation* occurs when both change in the same direction at the same time; *negative correlation* occurs when they change in opposite directions. Since the 1920s, pundits have espoused a belief that stock prices are positively correlated with hemlines.*

Correlation is not causation. Rather, correlation is a way of quantifying the co-occurrence of phenomena. Whether rising hemlines *cause* stock prices to increase (or vice versa) cannot be answered by correlation alone. At the same time, we can use the existence of a correlation to make forecasts: a "Hemline Fund" could buy and sell stocks based on the latest fashion trends and provided the correlation between hemlines and stock values actually exists, such a fund would do well.

To measure the correlation, imagine that our random variables were completely *nonrandom*. The simplest possible case is that they have a constant value: hemlines would always be 18 inches or that the share price of Enron would always be $80. This allows us to view the actual values as deviations from these fixed values. For example, if hemline lengths over a 5-year period were 16, 17, 20, 22, 15 inches, then the deviations would be −2, −1, +2, +4, −3. Meanwhile, if Enron share prices over the same period averaged $75, $85, $90, $40, $1, then the deviations would be −5, +5, +10, −40, −79.

These deviations can be treated as vectors: the hemline lengths correspond to the vector [−2, −1, 2, 4, −3], while the Enron share prices correspond to the vectors [−5, 5, 10, −40, −79]. We can then ask the question, *How similar are the deviations?* We can answer this using the L2-normalized dot product: this gives us the correlation, where values near 0 indicate little correlation, and values near 1 or −1 indicate a strong positive or negative correlation.

To calculate the correlation between different occupations, TORQ uses the deviation from the mean (average) of the respective vector components. In our example, the components of the dancer's ability importance vector [4.13, 2.25, 2.38] have a mean value of 2.92. This

*The initial observation is attributed to George Taylor, of the Wharton School of Economics. Taylor went on to be the primary architect of a New York law that allows public employees to form unions, though an amendment to the law makes it illegal for workers in those unions to go on strike.

produces a vector of deviations of $[1.21, -0.67, -0.54]$. Meanwhile, the components of the pile driver operator's importance vector $[4.25, 2.5, 2.62]$ have an average value of 3.12, giving a deviation vector of $[1.13, -0.62, -0.50]$. Finding the dot products of the L2-normalized vectors gives us a correlation of 0.9999999, indicating that the dancer's abilities are very closely correlated to the pile driver's abilities. In contrast, the correlation between dancer and motorcycle mechanic is only 0.1535, indicating very little relationship between the two sets of abilities.

On one level, this suggests that pile drivers and dancers have very similar knowledge and skill requirements. More generally, a person's background can be converted into an importance and level vector for a variety of skills, and so jobs suitable for a person's skills and backgrounds can be found easily.

However, while correlations are nondirectional, career changes are not: a dancer might try to become a pile driver or a pile driver might try to become a dancer. In either case, there will be some gaps in the person's skills, knowledge, and abilities. To compute this, TORQ uses a *Cross CMD*, which is the product of the importance value of the "to" occupation with the level value of the "from" occupation. This reflects the idea that a person currently in a profession presumably has a proficiency corresponding to the level, while the profession he or she is transferring into deems the skill, ability, or knowledge domain to have a specified importance. The difference between the Cross CMD and the CMD of the profession transferred into is designated the *gap*.

For example, suppose a pile driver wished to become a dancer. We compute the Cross CMD:

- The first component, for multilimb coordination, will be the product of the pile driver's level with the dancer's importance: $4 \times 4.13 = 16.52$.
- The second component, for visual color discrimination, will be $1.88 \times 2.25 \approx 4.23$.
- The third component, for hearing sensitivity, will be $2.38 \times 2.38 \approx 5.66$.

This gives a Cross CMD vector of $[16.52, 4.23, 5.66]$. The CMD for dancers (the "to" profession) is $[20.15, 4.23, 5.66]$. We can interpret our results as follows: while the pile driver's visual color discrimination and hearing sensitivity are at about the same level as those of dancers, his or her multilimb coordination is a little short (16.52 vs. 20.15).

What about a dancer who hopes to become a pile driver? In this case, the Cross CMD vector will be [20.74, 4.7, 6.24]. In this case, the Cross CMD components all exceed the pile driver's CMD components, suggesting that in some sense, the dancer already has the core abilities required of a pile driver.* In contrast, the pile driver hoping to become a dancer falls somewhat short in certain requisite abilities.

Lifelong Learning

To make the transition, some retraining will be necessary. Actually, this is true for entry into any occupation, so a perennial debate over education policy centers around how to prepare students for the jobs available in the future. But predicting the future is fraught with challenges, so these debates often devolve into contentious arguments and pointless recriminations, and the change occurs so slowly that by the time changes are implemented, new skills are required. Nevertheless, there are some long-term historical trends that should be taken into consideration.

First, contrary to popular narrative, the absolute *number* of jobs that don't require a high school diploma has remained more or less steady over the past 40 years.[†] However, these jobs are concentrated in fewer and fewer *occupations*. In 1973, about one-third of the occupations listed in O*NET required less than a high school diploma. By 1992, that fraction had dropped to 10% and, by 2016, less than 5% of occupations can be done by someone with less than a high school education. Consequently, if changing job conditions affect even a single one of these jobs, they will have a disproportionate effect on this group of poorly educated workers.

Mathematics has become increasingly important. In 2002, the survey of job analysts and incumbents concluded that about 13% of the occupations required a mathematical knowledge level of 4 or higher, corresponding to the ability to analyze data to determine the areas with

---

*This does not mean that the dancer can operate a pile driving rig! Rather, it means that the dancer is a good candidate for training to become a pile driver, because they have characteristics similar to those already in the profession.

[†]Carnevale, Smith, and Strohl, *Projections of Jobs and Education Requirements through 2018.*

the highest sales. By 2014, that level of knowledge was required in at least 30% of the occupations. Of even greater importance is that many occupations with minimal mathematical requirements in 2002 saw a substantial increase in those requirements by 2014. Of the 341 occupations that, in 2002, required only a level 2 of mathematical knowledge (just above that required to add two numbers), 151—nearly half—now require more. Consequently, a person whose mathematics education qualified them for a job in 2002 might be unprepared for the *same* job in 2014.

This last point is important, for most of the debate over educational policy concerns what should be taught in schools, from elementary to graduate (the so-called K–20 education). But even if someone completes a graduate degree, they will spend far more time in the workforce than they ever did at school; it is unrealistic to expect that what you learn in school will prepare you for your entire future career. Continuing education and lifelong learning are essential for maintaining relevance in the workforce. O*NET statistics reflect this: in 2002, a little over 9% of occupations required knowledge of education and training at level 4 or above. By 2014, that percentage had risen to more than 30%.

The Rise and Fall and Rise and Fall of Teaching Machines

One way to meet the ever-increasing need for education and training is to use teaching machines. There are three primary functions of a teacher: (1) to present information, (2) to evaluate the extent to which a student has learned the information, and (3) to adapt the instructional approach to allow the student to reach his or her learning goals.

Presenting information is often done by human beings, standing in front of a classroom giving a lecture. But 5,000 years ago, an unknown inventor created a revolutionary device to present information to a student: the book. More recently, other methods of presenting information have become widely available: audio and video lectures, on tape or online.*

*While these must be created by human beings, remember that the skills necessary may be very different: a superb human teacher might be a terrible writer and have no stage presence. Again, technology creates jobs at the same time it destroys them, but the skills required are often different.

Evaluating student learning is more difficult. For thousands of years, this has been accomplished by asking students questions and evaluating their responses. However, this requires questions with a clearly defined correct answer. We can ask, "Who was the seventeenth president of the United States?" and grade the responses, but "Should we legalize marijuana?" is much harder to evaluate, and while there is a correct answer to "What will the weather be on the day and place of your graduation?" we don't know it and can't evaluate an answer.

Even in cases where there is a clearly defined correct answer, it might not always be possible to identify when a learner has found it. For example, the seventeenth president of the United States was Andrew Johnson. But what if a student answers "Johnson"? As there are two presidents of the same name, this response is ambiguous; the respondent may answer "Johnson" because he is thinking of Gary Johnson (the 2016 Libertarian candidate for president) or perhaps even the basketball player Earvin "Magic" Johnson.

You might think mathematics, the epitome of objectivity, is immune to this problem, but it isn't. If we ask, "What is 3 + 2?" how should we respond to "2 + 3" or "1 + 1 + 1 + 1 + 1"? To a mathematician, both are equally correct responses to the question, and, depending on why we are asking the question, it's possible that these are more useful responses than the obvious answer "5." The problem is that we don't really want the answer to "What is 3 + 2?" What we really want is the answer to "What single digit number has the same value as the sum 3 + 2?"*

There are several ways to address this issue. A human teacher, given a technically correct answer that is different from what is expected, might prompt the learner for additional answers. If a teacher asked, "What is 3 + 2?" and a student answered, "Two more than three," a teacher might ask, "What do you mean by that?" and continue to request reinterpretation until the desired answer materialized.

Can machines mimic this process? One possibility is to reject all but the expected answer. However, this is a feeble imitation of what a

*One of my first real jobs was testing mathematics education software, to see whether it could accurately evaluate correct and incorrect answers. Given a question like, "What is 3 + 2?" a normal person would submit answers such as "5" or "32." I submitted the answer "3 + 2." The software accepted this as a correct answer to the question, and the company made some quick changes before releasing the module.

human teacher does. The problem is that in order for continued questioning to be useful, the incorrect answers must have at least some relationship to the desired answer. A student who answered, "What is 3 + 2?" with "Two more than three" could, under continued requests for clarification, arrive at the correct answer. In contrast, the student who answered "32" demonstrates a fundamental misunderstanding that no amount of reinterpretation will remedy.

The problem is that there are an infinite number of possible answers to the question. A human teacher can flexibly evaluate any response, while a machine is limited to its built-in responses. Consequently, the only way for a machine to mimic this aspect of human teachers is to limit the answers a student might give: in other words, by presenting a multiple choice question.

One of the first such devices appeared in US Patent 588,371, granted on August 17, 1897, to George G. Altman of New York City. Altman's device consisted of a set of interchangeable disks with subject relevant questions and several possible answers, only one of which was correct. Through a clever system of linkages, a bell would sound when the user pressed the correct answer. Altman noted the device could be used for teaching arithmetic or objects (i.e., vocabulary); more generally, it could be adapted to any subject where it is possible to ask questions with unambiguous "most correct" answers.

One of the shortcomings of Altman's device is that a user could find the correct answer by trial and error, pressing answers until the bell rang. Just because a user could get the bell to ring does not mean that they have learned something. Altman had the misfortune to invent his device just before a revolution in educational psychology made it and similar devices obsolete.

In the late 1890s, Edward L. Thorndike (1874–1949) ran a series of experiments on animal learning that began the *behaviorist school* of psychology. He placed cats in a cage, with a door that could be opened if the cat pressed a button in the cage. Initially, a cat would wander aimlessly in the box, but sooner or later it would accidentally step on the button and open the door, allowing it to escape. When the cat was returned to the box, it would again wander until accidentally stepping on the button, opening the door and allowing it to escape. By recording the amount of time it took between when the cat was placed in the box and when it triggered the button that opened the door, Thorndike produced a *learning curve*. As might be expected, the time required before the cat pressed the button decreased.

An ordinary person observing the process might conclude that the cat learned "Pressing the button opens the door." But Thorndike was not an ordinary person. He was a psychologist. Thorndike reasoned as follows: Suppose the cat learned "Pressing the button opens the door." Then there would be one last trial *before* the cat learned it, in which its time to escape would be measurable. On the next trial, with the insight "Pressing the button opens the door," that cat would be able to escape almost instantaneously; consequently, the learning curve would show a sharp and sudden drop in the time to escape, which would be maintained in future trials. However, the learning curves did *not* show such a drop; instead, they showed a more or less gradual decrease in the time required to escape. In particular, merely exhibiting behavior that *could* be explained by learning a particular concept is no evidence that the concept is actually learned; in this case, what the cat learned was not "Pushing a button opens the door," but "Random actions open the door."

How can this be applied to human beings? There are, even today, toys that claim to promote the learning of spelling or mathematics. These consist of puzzle pieces that, when fit together, spell out a word or a mathematical statement; misspelled words or incorrect mathematical statements correspond to puzzle pieces that cannot be fit together. An adult who knew how to spell would be able to assemble the puzzle pieces correctly, so we might conclude that anyone who could fit the pieces together knew how to spell. However, there are alternate routes to success: one might simply be good at fitting puzzle pieces together. For the task to be a true test of learning, these alternate routes to success must be closed.

This was the innovation of US Patent 1,050,327, given to Herbert Austin Aikins of Cleveland, Ohio, on January 14, 1913. Aikins was a psychologist and based his device on the studies of Thorndike and the behaviorists. Like the puzzles, his device required users to fit pieces together. However, the shapes of the pieces were concealed by a screen, so that the user could not use this information to solve the puzzle and would instead have to rely on their ability to solve the presented problem.

Further steps toward the "industrial revolution in education" were taken by Sidney L. Pressey of Columbus, Ohio. On May 22, 1928, Pressey received US Patent 1,670,480 for a "machine for intelligence tests."

Pressey's device delivered questions with multiple choice answers. The machine could be set to a testing mode, where only one answer would be permitted or a teaching mode, where a user could not advance to the next question until the current question had been answered correctly. Pressey had great hopes that the machine would launch a revolution in education. Unfortunately, Pressey's invention appeared on the eve of the Great Depression. There was little interest in replacing cheap human labor with expensive machines, so Pressey's invention was a financial failure, and by 1932, he had given up working on the device.

In 1953, educational psychologist Burrhus Frederic (B. F.) Skinner (1904–1990) visited his daughter's fourth-grade arithmetic class and made two disquieting observations. First, all students were being taught the same material at the same time. And second, there was a 24-hour delay before students learned whether or not their responses to a question were correct. Within a few days, he built a machine to address these concerns and, in the summer of 1954, presented an improved model at a conference at the University of Pittsburgh. At the time, Skinner was unaware of Pressey's work; however, shortly after an article on Skinner's machine appeared in *Science News Letter*, Pressey contacted him, and the two met that fall in New York City.

While Skinner acknowledged the importance of Pressey's machine, he also recognized that it was primarily a *testing* machine. By providing the machine with a mechanism for presenting new material, Skinner created the first actual teaching machine, receiving US Patent 2,846,779 on August 12, 1958 (assigned to IBM).

Skinner's machine relied on two important features. First, the material to be learned was broken into short, discrete pieces; Skinner introduced the term *programmed learning* for this approach. Second, Skinner preferred to avoid the multiple choice question, feeling that it was more useful to learn how to answer a question than to select the right answer from a collection of answers. Thus, in Skinner's machines, the student would write down an answer and then slide a panel to reveal the correct answer and to compare the two responses.

Skinner's device was a commercial success, though like Pressey's device, failed to bring about the industrial revolution in education. Skinner attributed his failure to cultural inertia: America of the 1960s was not interested in replacing human teachers with machines.

It's easy to blame cultural inertia, but there were several objective problems with 1960s-era teaching machines. First, one critic noted that whatever the value of programmed learning, the machines themselves were merely "expensive page turners." Second, Skinner's machine required the learner to self-evaluate, which risks giving the learner a false sense of their achievement. In response to "Who started the Protestant Reformation?" a student might write "Martin Luther King Jr.," then slide the panel to see the correct answer "Martin Luther." The student could very easily conclude that "Martin Luther King Jr." and "Martin Luther" were the same person and gain a false confidence of his knowledge of history.

Most important, the machines had very limited ability to adapt to student learning. If a student failed to learn one section, the machine could only present the same section in the same way a second time. Even if society were ready for teaching machines, the machines themselves were unready for use, and within a few years, teaching machines fell by the wayside once again.

Making the Grade

The promise of the industrial revolution is to free human beings from drudgery: tedious, repetitive tasks that require none of the qualities that make each person a unique individual. In teaching, there are no tasks more tedious than grading, and to some extent, the industrial revolution in teaching has fulfilled that promise by the invention of the multiple choice test and its relative, the true/false question. Thorndike was partially responsible: after his initial studies, he went on to design multiple choice intelligence tests for the US Army.*

There are numerous problems with multiple choice tests. We might take Thorndike's observation to heart: that someone can select the correct answer to a question does not mean that they know how to answer the question.

Consider: "Find 47 + 38." Obviously, one of the choices must correspond to the correct answer: 85. The other choices, known as *distractors*, are based on *error patterns*: common mistakes that occur when answering a question. One distractor might be 715, produced because the

---

*These were the precursors to the modern Armed Services Vocational Aptitude Battery (ASVAB).

person added 4 + 3 = 7 and 7 + 8 = 15. Another might be 22, from 4 + 7 + 3 + 8 = 22; an Abbott and Costello skit is based on this misunderstanding.

Now consider a person who does not know how to add but has a good number sense (the ability to accurately judge magnitudes). Such a person might conclude that 47 + 38 is "about" 100, and given a choice between the answers 22, 85, and 715, they could immediately eliminate 715 as too large, 22 as to small, and by process of elimination, conclude that 85 is the correct response *without* being able to add.

The obvious solution is to ask many questions. A person who knew addition would be able to answer these questions correctly; a person who relied on strategies other than addition would be unable to produce as many correct answers. But how many questions do we need to ask?

It depends on why we are testing. Educational researchers distinguish between *summative assessments*, which are designed to determine how much someone has learned, and *formative assessments*, which are designed to provide guidance for how to proceed. In general, summative assessments, as the name indicates, are done at the end of the learning period and are "high stakes," insofar as success or failure in a summative assessment is used to gauge whether someone has learned the material. By their nature, such tests must be standardized; hence, we have the SAT, GMAT, and state subject tests. In addition, there are professional tests, such as those needed to become certified as a physician or a lawyer. Clearly, it's in society's best interest to have some objective measure of how much a physician or lawyer actually knows.

However, if we accept the premise that learning is a lifelong process, then summative assessment is far less important than formative assessment. Because the goal of formative assessment is to guide student learning, and each student learns at a different pace, it's both unnecessary and counterproductive to standardize formative assessments. Thus, it might be worthwhile to produce a standardized 20-question test for summative assessment. But if the student has not yet mastered the learning objective, he or she will be likely to answer the first few questions incorrectly, and asking additional questions that are also likely to be answered incorrectly would be frustrating for the student and provide no extra information to the examiner. Similarly, if the student has mastered the learning objective, correct answers on the first few questions might be enough to indicate this and allow the student to move on to the next learning objective.

How many questions should we ask before we can draw a conclusion on the student's achievement? One approach is presented in US Patent 7,828,552, granted on November 9, 2010, to the Educational Testing Service (ETS) and developed by Valerie Shute (a cognitive psychologist, now at Florida State University), Erik Hansen (instructional psychologist still with ETS), and Russell Almond (a statistician, now at Florida State University). Their approach computes the *expected weight of evidence* for a set of questions (an approach championed by Almond for educational testing) and presents the questions with the highest expected weight.

In general, a test is a way to determine the presence of an unobservable quantity (student knowledge) from an observable quantity (number of questions answered correctly). Consider a cohort of 100 students, where 20 have learned arithmetic and 80 have not (we'll leave aside the question of what it means to have "learned arithmetic"). The *prior odds ratio*, the ratio of those who have learned arithmetic to those who have not, is 20 to 80, or 1 to 4.

Now suppose we have each student answer one question on arithmetic. Those who have learned arithmetic will almost certainly be able to answer the question correctly. In addition, if the question is multiple choice, some fraction of those who have not learned arithmetic will be able to select the correct answer by guessing. For convenience, suppose there are four choices, and that 1/4 of those who guess will guess the correct answer. This means we'll have 1/4 of 80, or 20 students, who don't know arithmetic but guess the correct answer. Altogether, there will be 40 students who answer the question correctly: 20 students who know arithmetic, and 20 who don't but got lucky in their guesses.*

Because we can't see into a student's head and determine whether they know arithmetic, we must base our decision of their knowledge level on their test result. "Obviously," the 60 students who failed to answer the question correctly don't know arithmetic (though we'll reexamine this assumption later). But what of the 40 who did answer the question correctly? The group of students who answered the question

---

*It's possible to try and correct for students who guess. For years, ETS penalized students for guessing on the SAT by subtracting additional fractions of a point for an incorrect answer. ETS discontinued this practice in 2016. We do not know if there is any connection between this decision and the ETS patent under discussion.

correctly has a *posterior odds ratio*, comparing those who've learned arithmetic to those who haven't, of 20 to 20, or 1 to 1.

Note the effect of the test: Without the test results, the best we could say is that the odds that a student knows arithmetic is 1 to 4 (hence, *prior odds*). Once we have the test result, then the odds that a student who passed the test knows arithmetic is 1 to 1 (hence, 1 *posterior odds*). The odds have increased by a factor of 4: we say that the *likelihood ratio* is 4. If our goal is to use the question to distinguish between those who know the subject and those who do not, we should choose questions with the highest likelihood ratios.

It might seem that determining the likelihood ratio requires finding the fraction of students who know the material. However, we don't actually need to know this. To see why, consider a different cohort, where 80 of 100 students have learned arithmetic. For this cohort, the prior odds of arithmetic knowledge will be 80 to 20, or 4 to 1.

What about the posterior odds? As before, correct answers would come from two groups of students: those who know how to add (80 students), plus 1/4 of those who don't but guessed correctly (5 students). Among the students who answered the question correctly, the posterior odds of arithmetic knowledge will be 80 to 5, or 16 to 1. Once again, our odds have increased by a factor of 4. In fact, regardless of how many students know the material, the likelihood ratio will be the same. If the prior odds are 10 to 1, then the posterior odds will be 40 to 1, and if the prior odds are 1 to 100, the posterior odds will be 4 to 100 (1 to 25).

This means that the likelihood ratio characterizes the *question* and not the students being tested. Moreover, it's not a coincidence that the likelihood ratio is 4, and that 1/4 of the students without knowledge can get the correct answer. As a general rule, if a question is answered correctly by everyone who has the knowledge the question tests for, and $\frac{1}{k}$ of those who don't have the knowledge the question tests for, the likelihood ratio for a correct answer will be k.

Our work thus far assumes that students who know arithmetic will get the correct answer. But as we all know, it's possible to know something and still answer a question about the topic incorrectly. This requires a minor modification of our approach. As before, suppose 80 of the students know arithmetic, and that 1/4 of those who don't know arithmetic will guess the correct answer. But this time, suppose that 1/4 of those who *do* know arithmetic will get the wrong answer.

In this case, 60 of the 80 students who know arithmetic will get the correct answer. At the same time, 5 of the 20 students who *don't* know arithmetic will also get the correct answer. The prior odds that a student knows arithmetic will be 80 to 20, or 4 to 1. However, if we focus on those who answered the question correctly, then the odds that a student who answered the question correctly knows arithmetic will be 60 to 5, or 12 to 1, giving us a likelihood ratio of 3. As before, the value of the likelihood ratio is independent of the fraction of students with knowledge. In general, if a question will be answered correctly by fraction p of those with knowledge, and fraction q of those without, then the likelihood ratio for the question will be p/q. In this case, 75% of those with knowledge, and 25% of those without, would answer the question correctly, giving us the likelihood ratio of 75%/25% = 3.

Statisticians define the *weight of evidence* to be the sum of the logs of the likelihood ratios. While it's traditional to use the so-called natural logs, the actual type of logs doesn't matter, and for illustrative purposes, we can continue to use our logs to base 2. A multiple choice question with 4 answers, where a correct answer will be given by all of those who know the material and 1/4 of those who don't, will have a likelihood ratio of 4 and a weight of evidence of 2.

The preceding analysis only distinguishes between those who possess the requisite knowledge and those who do not. But knowledge is rarely all-or-nothing: someone may be able to multiply 15 × 12 but have much more difficulty with 78 × 49. We can assess this level of proficiency in a similar fashion.

This requires some information about how those with knowledge will answer the question. Suppose we had a question with three answers A (the correct answer) and distractors B and C, which reflect common errors made by those who possess the requisite knowledge though not at the highest level of proficiency. If those who had the requisite knowledge gave the correct answer (A) 70% of the time, then those with knowledge but at a lower level of proficiency might make errors that lead them to answer B 20% of the time and to answer C 10% of the time. In contrast, suppose that those who don't have the knowledge will guess each of the three answers equally, so they will be correct 33% of the time and wrong 67% of the time.

Suppose a person selected answer A. The likelihood ratio will be 70% divided by 33%, or about 2.1. We can interpret this as follows: Whatever the prior odds that a person knew the subject, the posterior odds will be 2.1 times as great if they selected answer A.

What about answers B and C? Since a person who knows the subject answered B 20% of the time, the likelihood ratio for this choice will be 20% divided by 33%, or 0.6; likewise, for answer C, the likelihood ratio will be 10% divided by 33%, or 0.3. The posterior odds for these answers will be 0.6 and 0.3, respectively. Anyone who selects these answers will actually be less likely to know the subject.

To compute the likelihood ratio for the *question*, we weight the likelihood ratios. The logs of the likelihood ratios 2.1, 0.6, and 0.3 will be approximately 1.07, −0.74, and −1.74, respectively. Then, since 70% of those with knowledge will give answer A (and thus be "eligible" for the 2.1 likelihood ratio), 20% will answer B, and 10% will answer C, then our *expected weight of evidence* will be

$$0.7 \times 1.07 + 0.2 \times (-0.74) + 0.1 \times (-1.74) \approx 0.428$$

How shall we interpret these values? Remember that the log (to base 2) of a number is how many times we must double 1 before getting the number. In this case, 0.428 can be viewed as an average log, which means that the average odds ratio will be 1, doubled 0.428 times. Of course, we can't double a number a fractional number of times, but we can interpolate: doubling 1 zero times gives us 1, while doubling 1 one time gives us 2; doubling 0.428 times will give us something between 1 and 2. We can find a more exact value using calculus: it's around 1.346. If we present this question to a group of people who have the requisite knowledge, "on average," a correct answer to this question will increase the odds the student is one of those with knowledge by a factor of 1.346.

What can we do with this information? Consider the problem of trying to determine whether a student has knowledge. We can ask easy questions or hard questions. A question where only 70% of those *with* knowledge answer correctly might be classified as a hard question, while a question where 90% of those with knowledge answer correctly might be classified as an easy question.

What happens if we ask an easy question? For example, suppose we ask a question where a person with knowledge will give the correct answer A 90% of the time and the incorrect answers B and C 5% of the time (each). Again, if the persons without knowledge guess randomly and answer A, B, and C 1/3 of the time each, the odds ratios will be

90%/33% = 2.7 for A and 5%/33% = 0.15 for B and C, with logs to base 2 of
1.433, –2.74, and –2.74, respectively, and expected weight of evidence of

$$0.7 \times 1.433 + 0.05 \times (-2.74) + 0.05 \times (-2.74) \approx 1.016$$

This question has an average likelihood ratio of a little more than 2. The
likelihood ratio is *greater* than it is for the hard question.

This result may seem counterintuitive since it suggests knowledge is
more reliably demonstrated by answering *easy* questions. But suppose
you're interested in assessing the physical fitness of a cohort of students. One way is to present them with a challenging task: for example,
running a 100-mile race. While it's true that those who complete such a
task will be in good shape, we have no knowledge of the fitness of those
who fail to complete the task. It would be more informative to run the
cohort through a sequence of simpler exercises.

The Education of a Slave

Of course, unless the test results guide further instruction, formative
and summative assessments are indistinguishable. A student's final
grade in a class is a summative assessment of his or her work, but, at
the same time, the grade often determines which course the student
takes next. In general, any assessment is formative if the answers determine the learning topic that follows.

This is the heart and soul of the *Socratic method* of teaching. The
method, attributed to Socrates by his student Plato (427-347 BC), consists of the teacher asking a sequence of leading questions that cause
the student to discover knowledge. A good example is in Plato's dialogue
Meno, in which a slave, a person without any education, is led to discover the Pythagorean theorem through careful questioning.

To see how this might work, suppose we want to teach someone to
add 47 + 38. We might ask the following sequence of questions:

1. What's 40 + 7?
2. What's 30 + 8?
3. What's 4 + 3?
4. What's 40 + 30?
5. True or False: The order in which we perform an addition
   doesn't matter.

6. What's 40 + 7 + 30?
7. What's 7 + 8?
8. What's 70 + 15?
9. What's 47 + 38?

This sequence of questions is designed to elicit the insight that 47 + 38 can be found by decomposing 47 = 40 + 7, 38 = 30 + 8, and then recognizing that the value of a sum does not depend on the order in which we perform the addition. Thus, 47 + 38 can be found by adding 40 + 30 + 7 + 8; this, in turn, is 70 + 15, or 85.

This might seem a roundabout way of adding, and critics of this approach argue that for generations, American children have learned a quick, efficient algorithm for addition of multidigit numbers. While this is true, learning the algorithm is not the same as learning addition, any more than a cat's ability to escape a box is evidence that the cat learned how to escape the box. To understand this distinction, consider the problem of adding $(4x + 7) + (3x + 8)$. Many of those proficient in the standard algorithm for multidigit addition would be baffled by this problem and could only solve it after being taught a separate algorithm for the addition of algebraic terms (and, for generations, American children *have* learned it this way); however, those who learned *addition* as opposed to an algorithm would see this as not essentially different from 47 + 38.

We might go further. Because the standard algorithm for addition mechanizes the process of addition, it is a task that can be done more quickly, efficiently, and accurately by machines. Indeed, at the dawn of the twentieth century, *computer* and *calculator* were job titles, reserved for those who could perform calculations; by the end of the twentieth century, these jobs had all vanished. While there is some value to knowing how to compute without a calculator, a high level of proficiency is not necessarily a useful job skill: the real lesson of the story of the "Steel Drivin' Man" John Henry is that, in any competition of strength and speed between man and machine, the human loses. The only way to win is to transcend the machine; and the best way to do that is through continuing education.

6

\\\\\\\\\\\\\\\\\\

Forget Your Password?
Forget Your Password!

The new system will now require each password to consist of an
uncommon adjective, an occupation, some form of capitalization or
punctuation, a prime number, the user's favorite extinct animal
and how many yards Kenny Guiton threw for in last week's game.

CORY FRAME, SEPTEMBER 25, 2013
The Lantern (student newspaper of The Ohio State University, Columbus)

Since biblical times, stories abound over the use of *shibboleths:* words
that solve the *authentication problem* and verify your identity. These
range from the word *shibboleth*, used in the Bible to distinguish between
the Gileadites and the defeated Ephraimites; to *schilt ende vriend*, "shield
and friend," during a Belgian revolt against France in 1302; to *perejil*,
allegedly used by Dominican troops to decide who to murder during the
Parsley Massacre of 1937; to *lollapalooza*, in the Pacific theater during
World War II.

Today's most common shibboleth is the password. However, the authentication problem has changed. In previous eras, those who failed
the authentication process faced massacre or imprisonment, and so
the problem was making sure that you passed it. Today, the problem
is making sure that *only* you can pass the test. A hacker who finds your
password has access to everything.

Suppose user Alice sets up an account and chooses the password "rabbit." Eve, a hacker, tries to gain access to Alice's account. To do so, she submits Alice's user name and is prompted to enter Alice's password. Eve tries common passwords: lists are published online, both by legitimate companies seeking to improve computer security, and hackers seeking to help other hackers. Thus Eve might enter "123456" and fail to log in; try "password" and fail to log in; try "qwerty" and fail to log in.

One way to make Eve's task more difficult is to freeze the account after a number of failed log-in attempts. Eve might only have three tries to guess Alice's password, and if she fails to do so, she will be unable to make more attempts. This might seem to make password hacking nearly impossible, but there's a weak point: the *system* needs to know the password. In the early days, this was done by maintaining a password file. Alice submitted her name and the password "rabbit"; a computer would look up Alice's name and confirm that her password was "rabbit" and give her access to the system. Since the passwords are stored in a computer file, then anyone who stole the computer file would have *every* user's password.

Remarkably, it took several years (and several incidents of stolen password files) before the obvious solution was presented: encrypt the password file. This was first suggested in 1968, by Maurice Wilkes (1913–2010), an English computer scientist at Cambridge University. We'll deal with the details of modern encryption in a later chapter; for now, suppose we encrypt our passwords by using a *shift cipher* that replaces each letter with the letter after it (with "z" replaced by "a"). Then Alice's password "rabbit" is encrypted as "sbccju."

There are some advantages to maintaining an encrypted file of passwords. First, even if the password file were copied, the data would be protected from prying eyes: Alice's password is not "sbccju." Second, if a user ever forgets his or her password, a system administrator can look it up and decrypt it. If Alice forgot her password, the administrator could look up the encrypted form "sbccju" and decrypt it as "rabbit."

As an alternative, we can use a *hash function*: a procedure that chops and mixes data to produce a *hash value*. A simple example of a hash function is alphabetizing the letters of a word or phrase. If Alice's password is "rabbit," the hashed value would be "abbirt," and the hashed value, not the password itself, would be stored. The advantage is that the hash file can't be decrypted, so even if a hacker knew every detail of the hash function, the hash file would still be useless for recovering passwords. At the same time, this is a disadvantage: if a legitimate user forgets his or her password, the system administrator can't recover it, and so the user must choose a new password.*

Now suppose Alice submits her password. Her submitted password would be processed in the same way as her original password, and the encrypted or hashed versions of the submitted password would be compared to the encrypted or hashed versions of her real password. If she submitted "rabbit," the system would replace each letter with the one after it, to produce "sbccju," which matches the encrypted version of Alice's password. Or the system would alphabet hash the letters to produce "abbirt," which matches the alphabet hash version of Alice's password.

Suppose a hacker steals the file. Since they don't have the actual passwords, it might seem they can't recover them.† However, they can try a *dictionary attack*: take a dictionary of common passwords, encrypt or hash them as the system would, and then check the file to see if there are any matches. If "rabbit" and "123456" are in the dictionary, then the person who used the former as a password is no more secure than the person who used the latter.

Keyspace

A dictionary attack is an example of a *brute force attack*, which tries every possibility. The set of all possibilities is known as the *keyspace*. Intuitively, the bigger the keyspace, the more difficult a brute force attack and the more secure the password.

How big is the password keyspace? Most passwords consist of *strings*

---

*A good way to determine whether a site hashes or encrypts passwords is how it responds to a "Forgot my password" notification. If the site sends an email with your password, the password is stored somewhere, and amenable to being hacked. On the other hand, if the site resets your password, the password is probably hashed.

†For the shift cipher and alphabet hash, it's easy enough for an attacker to find passwords that work. However, real ciphers and hash functions are much more sophisticated, and it's believed that, in practice, it is impossible to break the encryption or hashing of a password file.

(sequences) of *characters*, where the individual characters can be typed using some input device, typically a keyboard. There are 95 such characters on all keyboards: 62 *alphanumeric characters*, corresponding to digits or uppercase or lowercase letters of the alphabet; and 33 *non-alphanumeric characters*, corresponding to punctuation marks and other symbols: #, @, !, and so on.

In practice, a site sets a *password policy*, and users follow the guidelines of the policy to produce their password. For example, consider a site that sets a policy "choose a single character as your password." User passwords will be single characters such as "X," "#," or ">." This policy produces a keyspace containing 95 elements.

What if the password policy required choosing several characters? To determine the size of the keyspace, we can use the *fundamental counting principle*: Suppose you have a sequence of choices to make. If you have a ways of making the first choice and b ways of making the second choice, you have $a \times b$ ways of making the two choices. Thus, if the password policy was "choose a two-character string as your password," then there are 95 choices for the first character of the string, and 95 choices for the second character of the string, so there are $95 \times 95 = 9,025$ ways of choosing a two-character string, and our keyspace will have 9,025 elements. Likewise, if the password policy was "choose a six-character string," our keyspace would have $95 \times 95 \times 95 \times 95 \times 95 \times 95 = 735,091,890,625$ different elements.

It's imperative that sites set a password policy, for if they don't, users will gravitate toward the simplest possible passwords. In 1979, Robert Morris and Ken Thompson, computer scientists at Bell Laboratories, studied 3,289 passwords gathered from a group of users. They found

- 15 consisted of a single character,
- 72 consisted of two characters,
- 464 consisted of three characters,
- 477 consisted of four characters,
- 706 consisted of five characters,
- 605 consisted of six characters, and
- 492 could be found on commonly used name lists and dictionaries.

All of these would be considered bad passwords: about 86% of the total.

Of course, 1979 was the pre-dawn of the internet era, and surely we've grown in sophistication. In 2005, David Salomon, at California State University Northridge, conducted a similar survey of 6,828 passwords. He found

- 33 consisted of a single character,
- 151 consisted of two characters,
- 911 consisted of three characters,
- 946 consisted of four characters
- 1,353 consisted of five characters,
- 1,280 consisted of six characters, and
- 993 could be found on commonly used name lists and dictionaries.

In 26 years, the fraction of bad passwords had barely declined: fully 83% of passwords in the survey could be considered bad passwords.

Sites can implement password policies that prohibit the use of bad passwords. However, while a site's password policy might allow for a very large keyspace, most users select passwords from a very limited part of the keyspace. It's as if you could go anywhere in the world for vacation, but choose to spend it in your living room. Simply measuring the size of the keyspace will generally overestimate the real strength of a password.

One way to encourage users to select stronger passwords is through a *password strength meter*. One of the earliest appeared on March 23, 2010, when Netscape Communications received US Patent 7,685,431 for "System and Method for Determining Relative Strength and Relative Crackability of a User's Security Password in Real Time," invented by Netscape's director of product management, Michael Mullany. The patent is currently owned by Facebook.

The Netscape approach uses a common tactic in mathematics: inverting the question. Instead of using the password policy to create a password, we use the password to create the password policy. The challenge is that many different policies could produce the same password. For example, suppose a user's password is "Z". There are many different password policies that could have produced this password, and each has a differently sized keyspace:

- The policy "choose a capital letter" has keyspace 26.
- The policy "choose an alphanumeric character" has keyspace 62.
- The policy "choose a single character" has keyspace 95.

Assume that these are the only possible password policies. Which keyspace should we use? In general, an analysis of this sort relies on the *worst case scenario*. This isn't because we're pessimists; it's almost exactly the opposite: If we plan for the worst case scenario, then reality will be better.

If a user's password is Z, the worst case scenario corresponds to a password policy of "choose a capital letter."* A dictionary attack would only have to check 26 possibilities: the worst case scenario. But if the actual password policy was "choose a single character," then, in practice, a dictionary attack would have to check all 95 possible single character passwords: it's harder for the attacker to mount the dictionary attack, which is better for the user.

To evaluate the strength of a password, the Netscape application breaks the password into numerals, lowercase letters, uppercase letters, and non-alphanumeric characters. From this, a worst case scenario password policy can be constructed for passwords of any length. For example, consider the password "rabbit." This consists of 6 lowercase letters. The worst case scenario password policy is "choose 6 lowercase letters to form the password." The corresponding keyspace will be $26 \times 26 \times 26 \times 26 \times 26 \times 26 = 308,915,776$.

On the other hand, suppose the password was "Rabbit." This consists of a mix of uppercase and lowercase letters, so a worst case scenario password policy would be "choose 6 uppercase or lowercase letters to form the password." This gives us a keyspace of $52 \times 52 \times 52 \times 52 \times 52 \times 52 = 19,770,609,664$. And if the password was "Ra8b:t," then the worst case scenario password policy would be "choose 6 characters to form the password," giving us the largest possible keyspace for six characters, namely, $95 \times 95 \times 95 \times 95 \times 95 \times 95 = 735,091,890,625$. To convert this information into password strength, the Netscape approach uses informational entropy.

*We could, of course, create even smaller keyspaces: "Choose a letter in the second half of the alphabet" or even "Choose a letter from the word ZOO" or even "Choose the letter Z." Deciding on a keyspace is more of an art than a science, and so technically we're looking at the worst *plausible* case.

Consider the game of Twenty Questions, where one person thinks of an object and the other can ask up to 20 questions about it in an attempt to guess what it is. These questions are supposed to be yes/no questions (though we often begin with "Animal, vegetable, or mineral?" as an exceptional first question). If we imagine the password to be the object being sought, then the strength of a password can be measured by the average number of questions required to determine it.

For example, suppose your password is a three-digit sequence. There are 1,000 of these, from 000 to 999. A hacker *might* ask the question, "Is the password 000?" and, if the answer is yes, have the password immediately; on the other hand, the answer might be no, and they'd have to ask another question. A particularly dull hacker might follow this question with "Is the password 001?" and "Is the password 002?" and so on. In a worst case scenario, the password is 999, which would require the hacker to continue asking questions until they finally arrive at "Is the password 999?" This takes a total of 1,000 questions, and we might conclude the password is very secure. However, sequential guessing is a particularly inefficient way to obtain the number. Remember that information is obtained by reducing uncertainty; we'd like our questions to eliminate as much uncertainty as possible.

Initially, we have no information about the password, so we know it is one of the thousand possible numbers from 000 to 999. If we ask, "Is the password 000?" a yes answer would reduce the possibilities for the password down to one, namely, 000; this would reduce our uncertain by eliminating 999 other possibilities. But if the answer was no, then we'd have a far more modest reduction in uncertainty. Indeed, only one password possibility, namely, 000, would be eliminated, and 999 would remain.

Thus, we might instead ask, "Is the password 500 or more?" In this case, a yes answer will eliminate 500 possibilities (the passwords 000 through 499), while a no answer would also eliminate 500 possibilities (the passwords 500 through 999).

Suppose the answer to the question is no. This means the password is between 000 and 499. Our next question should eliminate as many of these as possible, regardless of whether we answer the question yes or no. Thus we might ask "Is the password 250 or more? A yes answer eliminates the 250 possibilities 000 through 249, while a no answer would eliminate the 250 possibilities 250 through 499." If we continue in this fashion, we find that we can guess a three-digit sequence in just 10 guesses. Thus the three-digit password has a strength of 10 bits: we can find the password by asking 10 questions.

One problem is that it might not be easy to see what question we should ask. To avoid this problem, we can proceed as follows. The most efficient yes/no question will separate our possibilities into two equal-sized collections of passwords. Every yes/no question we need to ask to determine a password doubles the size of our password space. Thus, if we ask 10 questions, our password space has $2 \times 2 \times 2 \times 2 \times 2 \times 2 \times 2 \times 2 \times 2 \times 2 = 1,024$ elements; consequently, our three-digit password (which comes from a password space of 1,000 elements) can be found by asking 10 or fewer *efficient* questions.

We can go further: guessing the number this way *always* takes 10 questions, regardless of what the number actually is. However, since passwords are chosen by people, we might use an insight on how people think to ask more efficient questions. For example, suppose that half the time, the chosen password is "777." A clever hacker might ask the question, "Is the password 777?"

At first glance, it might not be obvious why asking, "Is the password 777?" is better than asking, "Is the password 000?" In both cases, a no answer only eliminates one possible password. But there's a crucial difference. If the first question is, "Is the password 777?" then 50% of the time, the hacker guesses the password within one question. The other 50% of the time will require asking 10 more questions (which could be the same as those above, starting with "Is the password 500 or more?"). Thus on average, the hacker needs

$$1\,(0.50) + 11\,(0.50) = 6 \text{ Questions}$$

We can interpret this to mean that our password policy, where users choose a three-digit number to serve as their password, has a strength of 6 bits.

The preceding relies on our ability to frame the right set of questions. In lieu of this, we can compute the informational entropy. Imagine that we've divided the users into sets, where users in each set choose a password according to some rule that has a keyspace of k passwords. Then it's as if this set of users followed the password rule "choose one of the following k passwords." The informational entropy for this particular password policy is then $\log(k)$, whereas before, $\log(k)$ is the number of times we must double 1 before arriving at k. This informational entropy translates into the average number of questions we must ask one of these users before we can determine their password. If we do this for all users, we have partitioned the users into distinct sets, every one of which follows some password rule. We can then find the weighted average of the entropies.

For example, suppose our password policy requires a six-character password. Imagine that some users choose a password as follows: They write down each of the 95 symbols on a tile, put all the tiles into a bag, and then create a password one character at a time by drawing one tile from the bag, writing down the symbol on it, and then returning the tile to the bag. This group will produce passwords like "39xK!)" or "K0+-}z." Since there are 735,091,890,625 possible six-character passwords, the entropy for this group will be $\log(735{,}091{,}890{,}625) \approx 39.42$.

On the other hand, consider a different group of users who choose a six-letter English word. This group will have passwords like "zouave" or "sashes." There are perhaps 20,000 six-letters words in English. For this group, the password entropy will be $\log(20{,}000) \approx 14.29$.

Or consider another group of people: they pick a letter and repeat it six times to form their password. This group will produce passwords like "mmmmmm" or "AAAAAA." There are 52 such passwords, and so this group's password entropy will be $\log(52) \approx 5.70$.

Now consider real users. Suppose 80% of users chose passwords by picking a letter and repeating it six times; 19% chose passwords by picking a six-letter word; and 1% chose a six-character password by picking characters completely at random. As a whole, these users will have password entropy

$$5.70\,(0.80) + 14.29\,(0.19) + 39.42\,(0.01) \approx 7.67$$

On average, it will take a hacker 7.67 questions to find the password of a user. This value is known as the *guessing entropy*.

Notice that the general password policy, of choosing six characters, has a high entropy: 39.42 bits. However, this doesn't translate into a high guessing entropy, because very few users (1%, in our example) select six characters completely at random. In effect, it doesn't matter that users *could* choose any one of billions of possible passwords, since most users choose one of the 52 possible repeated letter combinations.

So what can we do? Paradoxically, we can make passwords stronger by making the keyspace *smaller*. In a simple example, suppose we forbid users to select a six-letter word or a single letter repeated six times as their password. This reduces the size of our keyspace, making it easier to mount a brute force attack. However, it also forces users to play in the rest of the field: the 80% of users who *would* have chosen a single letter, repeated, and the 19% of users who *would* have chosen a word, must now choose something else. Suppose 10% now choose letters at random.

What do the other 90% do? The National Institute for Standards and Technology notes in its guidelines for electronic security that rules that constrain password choices will "generally be satisfied in the simplest and most predictable manner." Thus, if repeated letters are forbidden, and words are forbidden, then users will gravitate toward the simplest nonrepeated, nonword sequences. One possibility is six letters in sequence: *abcdef, bcdefg*, and so on. There are 42 such sequences (counting capital letter sequences), so the group that does this will have password entropy $\log(42) \approx 5.39$, and our password policy now has guessing entropy

$$5.39\,(0.90) + 39.42\,(0.1) \approx 8.79 \text{ bits}$$

Thus, even though our keyspace is smaller, because we've eliminated many possible passwords, users are forced to use more of it, which increases our guessing entropy.

In practice, we can't compute the guessing entropy without detailed knowledge of how users choose passwords. However, the preceding illustrates a general result: If we forbid certain commonly used methods of generating passwords, we'll increase the guessing entropy. If we also forbid letters in sequence, then the users who would have chosen such passwords will be forced to choose other, more difficult, passwords.

Correct Horse Battery Staple

While the guessing entropy may be the best way to measure the strength of the password, it requires detailed knowledge of how all users of all systems choose passwords. As an alternative, Dan Lowe Wheeler of Dropbox released the *open source* program zxcvbn on April 10, 2012.

An open source program is one that is released into the wilds of the internet, whose coding is readable and modifiable by anyone. Typically, open source software is free, though this is not a requirement, and much *freeware* is not open source.* Open source has potential to transform the computer software industry: users have a choice between open source platforms like Open Office (which is what this manuscript was prepared with) or paying several hundred dollars for copies of Microsoft Office.

The legal status of open source software is unclear, in part because the US Patent and Trademark Office has been monumentally stupid in issuing patents. For example, one requirement is that patents must be for inventions that are nonobvious to "those skilled in the art." Consider the following problem: someone wants a web page to open up a new web page (a pop-up window, in other words). To those skilled in the art — in other words, to people who design web pages — this is a trivial problem that takes three lines of JavaScript to accomplish. Yet on October 30, 1998, Brian Shuster of Ideaflood applied for and eventually received US Patent 6,389,458, which covers any software implementation, past, present, or future, that opens up a new browser window. All such implementations now owe a royalty fee to the patent owner: Intellectual Ventures.

One potential problem with open source software is that if it's not patented, it's conceivable someone who is not the developer might apply for and receive a patent and prevent others from using it. On the same date he released zxcvbn, Wheeler also submitted a patent application for it. On March 3, 2015, he received US Patent 8,973,116. The system is currently used by Dropbox.

Suppose a company's password policy requires that passwords not be a word in any common dictionary and that passwords must include numbers and uppercase and lowercase letters. One way to produce such a password is to substitute numbers and symbols for similar-looking letters. Thus, the digits of the number 1337 look somewhat similar to the letters *l*, *e*, and *t*, and so 1337 can be read as the word *LEET*, which gives rise to the *leet alphabet*. There's no formalized leet alphabet, but common substitutions are 1 for *i* or *l*, 8 for *B*, 0 for *O*, 1< for *K*, and so on.

Suppose Alice wants to use the password "rabbit." Since this is a common word, it would be forbidden by the password policy, but she

*The critical difference is that since the code for open source software is known to everyone, there are no concerns that the software might be doing something other than what it's supposed to do.

could change it to "Ra88it," which conforms to the requirements that the password not be a word and include numbers and capital letters. This suggests that Alice's password could originate from the rule "choose six alphanumeric characters." There are $62 \times 62 \times 62 \times 62 \times 62 \times 62$ = 56,800,235,584, so the password entropy will be log(56,800,235,584) = 35.73 bits, evidently a good password.

However, *portions* of Alice's password can be described using low entropy rules. Dropbox (and patented version) relies on a collection of dictionaries: for example, commonly used words (e.g., *password*); names (e.g., *Smith, Johnson*), patterns (e.g., *zxcvbn*, formed by adjacent letters on the keyboard), and numbers (e.g., 7, 13, 69). We can break Alice's password apart into segments, and see if we can find each segment in one of our available dictionaries. If we can, then that portion of the password can be treated as being formed by the rule "choose an entry in this dictionary," and we can compute the corresponding entropy. If a word appears in several dictionaries, we'll use a worst case scenario approach and use the rule that gives us the lowest entropy.

For example, suppose we break Alice's password into three segments: *Ra, 88*, and *it*. We can make the following observations:

- *Ra* is a two-letter name (the Egyptian sun god). More generally, it's one of a class of two-letter words. If we view *Ra* as the name of the Egyptian sun god, then the corresponding password rule would be "choose the name of a god or goddess." As there are several thousand possibilities, this would have informational entropy of around 12. However, if we view "Ra" as a capitalized two-letter word, then it can be produced by the rule "choose a two-letter word with capital or lowercase letters." There are about 100 two-letter words, and with capitalization variations, this number increases to 400. Thus, *Ra* can be viewed as coming from a password policy with entropy log(400) $\approx$ 8.64. We use the lower value.

- *88* is a two-digit number. However, the digits are repeated, so that can be produced from the password policy "choose a digit and repeat it." There are 10 possible choices, so *88* can be viewed as coming from a password policy with entropy log(10) $\approx$ 3.32.

- *it* is a two-letter word. Since it isn't capitalized, it can be viewed as coming from a password rule "choose a two-letter word."

There are about 100 such words, so this last component could be produced by a password policy with entropy $\log(100) \approx 4.64$.

This breakdown allows us to view Alice's password as the result of *three* password policies, each producing a portion of her password.

Now imagine a hacker trying to guess Alice's password. Since the first portion was produced by a rule with entropy 8.64, the hacker would have to ask an average of 8.64 questions to find the first part of Alice's password. The second part, coming from a rule with entropy 3.32, would require an average of 3.32 questions to find. Finally, the third part, coming from a rule with entropy 4.64, would require an average of 4.64 questions to find. The hacker would be able to determine Alice's entire password by asking an average of 8.64 + 3.32 + 6.64 = 18.6 questions. Consequently, Alice's password can be viewed as coming from a password policy with an entropy of 18.6.

As above, there are many other policies that would have produced Alice's password. For example, we might view her password as the result of six password policies, one for each of the letters in her password. These would be as follows

- Choose a capital letter for the first character: entropy $\log(26) \approx 4.7$.
- Choose a lower case letter for the second character: entropy $\log(26) \approx 4.7$.
- Choose a number for the third character: entropy $\log(10) \approx 3.32$.
- Choose a number for the fourth character: entropy $\log(10) \approx 3.32$.
- Choose a lower case letter for the fifth character: entropy $\log(26) \approx 4.7$.
- Choose a lower case letter for the sixth character: entropy $\log(26) \approx 4.7$.

Alice's password might have been the result of these six policies, with total entropy

$$4.7 + 4.7 + 3.32 + 3.32 + 4.7 + 4.7 = 25.44 \text{ bits}$$

Other possible password policies could lead to the password *Ra88it*; we'll choose the worst case scenario and use the lowest entropy, corresponding to the fewest number of questions the attacker must answer to correctly guess Alice's password.

How secure is Alice's password? Again, in practice, someone trying to log in to a system as Alice would have a hard time, even if her password was *alicespassword*, because the hacker would have a limited number of tries before being locked out. But if someone were to steal the password file, a dictionary attack could succeed. With an entropy of 18.6 bits, Alice's password comes a keyspace with about 400,000 passwords, and a brute force attack could be completed in a matter of seconds. In short, our vaunted password policy has forced Alice to use a password that is difficult for her to remember and, at the same time, is easy to break.

Entropy is additive, so the longer the password, the more secure it is. Consider a pass phrase like "correct horse battery staple," presented by Randall Munroe, author of webcomic XKCD, as an example of a good password. The policy that produces such a password might be described as follows:

- Choose a common English word. There are perhaps 2,000 common English words, so this rule has entropy $\log(2,000) \approx 10.97$.
- Choose a second, third, and fourth English word. Each of these adds 10.97 more bits of entropy. The strength of the password "correct horse battery staple" will be about 44 bits, and it comes from a keyspace with 16,000,000,000,000 elements. A computer that could find "Ra88it" in a few seconds would take several years to find "correct horse battery staple."*

The preceding analysis suggests that there is a simple solution to the problem of user passwords: simply require them to be of a certain minimum length, with some sort of informational entropy measurement to ensure that the passwords aren't easily guessable like 12345-678987654321. Thus, a system might require the password be a pass *phrase* of a specified length.

---

*Some security consultants have criticized Munroe's approach, arguing that multiword dictionary passwords are also used, rendering such a password insecure. But you can't fight entropy: the keyspace that produces "correct horse battery staple" is 40 million times larger than that which produces "Ra88it," and the password is correspondingly harder to guess.

However, many systems limit the length of passwords.* Users will be confronted with policies that require them to choose less and less memorable strings as passwords. One possibility, suggested in 1980 by science fiction author Isaac Asimov (1920–1992), is for the user to choose a literary work known to him or her, and choose the initial letters of each line or word: for example, the first letter of the lines of a favorite sonnet (which would give a 14-letter password like "STRASAABBN-NWSS") or a well-known phrase ("WTPOTUSOA"). This produces long, seemingly random passwords with relatively high entropies.

The Fist of the Sender

Of course, it would only be a matter of time before dictionary attacks began identifying "STRASAABBNNWSS" or "WTPOTUSOA" as likely passwords, so Asimov's solution would be temporary at best, and unless we're able to use very long passphrases, users will be condemned to create passwords like "eH89!@nKs."

This leads to a new problem. While it will be harder for a hacker to guess a password like "eH89!@nKs," a user who "remembers" it by writing it down on a sticky note on their laptop becomes vulnerable to more traditional means of stealing data. This is a serious source of concern for larger organizations. For example, janitorial work is often done after hours, and everyone needs their trash emptied and floors vacuumed. Thus the custodial staff is generally able to enter offices unsupervised, and in theory has access to all sorts of secrets. One could simply bribe a janitor to pass on any sensitive information he or she came across at work: for example, by writing down any seemingly random phrases written on sticky notes next to computers.†

*At the dawn of the web, my email password was "Elgin marble 5," which (based on the preceding) has an informational entropy on the order of about 30 bits. But our email system migrated to a new system, which limited passwords to 8 characters, and I was forced to choose another password that could be more easily guessed.

†During World War II, when wartime espionage was of particular concern, illiteracy was often a requirement for janitorial jobs, under the assumption that an illiterate person could not distinguish between secret documents and mundane ones. While it's possible that an illiterate person could pass on a paper with an important wartime secret, it's more likely they would pass along a lunch order or the phone number of a repairman.

Because of this, there is a great deal of interest in alternatives to passwords. One much-touted possibility is the *biometric password*. Roughly speaking, a biometric password depends on some physical characteristic of the user, such as a fingerprint. However, biometric passwords fail to solve a more fundamental problem: *continuous verification*.

Suppose an authorized user logs on to a secure site but fails to log out before leaving for lunch or going home. Then whomever can physically access their computer will gain access to the protected site. Users of mobile phones have a similar problem: we might enter in a password once, but thereafter the device stores it and automatically logs us in to wherever we want. Anyone who steals our device will be able to access anything we could.

The obvious solution is to require the person to periodically resubmit their credentials. However, this has the effect of interrupting a user at constant intervals: driving anywhere would be difficult if you had to show your license every 10 miles. An ideal method of continuous verification would run in the background; for example, a cell phone could monitor how it was held and how the user walked. One possibility is to use *keyboard forensics*.

Keyboard forensics is based on a simple idea: using a keyboard is a physical task that relies on a complex interaction between brain and body, and so each person's use of a keyboard is slightly different. This idea predates the computer: early telegraph operators, sending signals via Morse code, could often identify each other by their keystroke signature: the so-called *fist of the sender*. For example, in Morse code, the letter *s* consists of three short bursts (dots), while the letter *e* consists of a single dot. Since dot-dot-dot-dot would be unrecognizable (it could be *s-e*, *e-s*, or even *e-e-e-e*), operators must include short pauses. A particularly cautious operator might include a long pause: dot-dot-dot, pause, pause, pause, dot. Another operator might include a barely perceptible pause between the signals: dot-dot-dot, dot. As late as World War II, this method would be used to identify telegraph operators and offer some assurance that the sender was in fact the person it claimed to be.

There are several important advantages to this method of identification. A person can be forced to divulge their password or passphrase (a so-called *rubber hose attack*, a reference to beating the victim with a rubber hose until they give their password, though the more gentle modern version is a *phishing attack*, where an "urgent message" from a secure site prompts the user to give away his or her user ID and password). On the other hand, a person can't explain their typing pattern. Moreover, if they are being coerced, stress is likely to change their typing pattern, or they could deliberately alter their typing pattern to thwart authentication.

For example, consider someone typing the phrase "The quick brown fox jumps over the lazy dog." This phrase consists of a sequence of characters. As human beings, we group them into words; but as data, we can group them into two character *digraphs*, three-character *trigraphs*, and so on. In this case, the phrase would break down into digraphs like *t-h*, *h-e*, *e*-space, space-*q*, and so on. Not all digraphs are equally easy to type. For example, a trained typist will likely type the digraph *a-z* more slowly than *l-a*, because the former requires typing the two letters with the same finger. In contrast, a hunt-and-peck typist might have actually be faster with *a-z*, because the two letters are adjacent on the keyboard.

In theory, this suggests that you can identify a typist by how quickly he or she types digraphs. In 1980, R. Stockton Gaines, William Lisowski, S. James Press, and Norman Shapiro, of RAND Corporation, tested this theory by giving seven professional typists a paragraph of prose to type. They recorded the times required to type each digraph of the text. Four months later, they gave the same paragraph to six of the typists and compared the intervals between the keystrokes. They found that the keystroke patterns of the typists formed a clear fingerprint that could be used to distinguish between them. In their study, they focused on five digraphs: *i-n*, *i-o*, *n-o*, *o-n*, and *u-l* (all digraphs typed with the right hand, which they acknowledged was a key limiting factor in their study).

Consider any one of these digraphs, say, *i-n*. In general, the time it takes a person to type this digraph will vary slightly. But since the digraph will likely appear many times in a long text (for example, it's appeared several times in the past few paragraphs), we can form the histogram of the times required to type the digraph.

Now suppose an authorized user leaves without logging out of a secure site, and an unauthorized user tries to use the system: for example, they want to send some emails using the original user's account. Again, the digraph *i-n* will generally appear many times, we can form a histogram of times required to type the digraph. We can then analyze the two histograms, to see if they're produced by the same person. There are several ways of measuring the difference between the two histograms (many of which were discussed in the chapter on image comparison); Gaines and his colleagues found that for certain digraphs, the difference was substantial enough to form the basis for user authentication.

On September 23, 2014, Identity Metrics received US Patent 8,843,754 for "Continuous User Identification and Situation Analysis with Identification of Anonymous Users through Behaviormetrics," developed by Herbert Lewis Alward, Timothy Erickson Meehan, James Joseph Straub III, Robert Michael Hust, Erik Watson Hutchinson, and Michael Patrick Schmidt.

The Identity Metrics patent relies on collecting *trigraph* timing: how long it takes the user to enter in a sequence of three symbols, like *s-t-r*. Imagine that, over time, we've produced the timing histogram for this trigraph for the authorized user. The histogram can be described in terms of a *probability density function*. The most familiar probability density function is the *normal distribution,* usually called the *bell curve*; however, there are an infinite number of other possibilities. The probability density function gives us an idea of how frequently an event occurs. For example, our probability density function might tell us that 5% of the time, it takes the user less than 100 milliseconds to type *s-t-r*; that 10% of the time, it takes the user between 100 and 150 milliseconds to type *s-t-r*; and so on.

Now imagine that *someone* is typing, and we collect this person's timing histogram for the trigram *s-t-r*. We can compare the two histograms to determine whether the typist is the authorized user. One simple possibility is to use the histogram difference, in the same way we compared whether two photographs were of the same object: if the difference between the timing histograms exceeded a certain threshold, the user would be locked out or, less obtrusively, be requested to reenter a password.

Hypothesis Testing and Likelihood Ratios

The Identity Metrics patent relies on a more sophisticated approach using *statistical inference.* To understand how this is done, consider the following problem: You have a coin, which might or might not be a fair coin. To determine whether the coin is fair, you'll flip it a number of times and observe the number of times it lands heads and the number of times it lands tails. If you flip the coin 100 times, and find it lands heads all 100 times, you'd probably conclude the coin is not fair. If the coin lands heads 50 times, you'd probably conclude the coin is fair.

Classical inferential statistics begins as follows. We have two *hypotheses*, one of which corresponds to the *true state of the world*. In this case, our first hypothesis is that the coin is fair; the second is that the coin is not fair. One of these hypotheses is known as the *null hypothesis*. Our evidence will then lead us to one of two conclusions: (1) we reject the null hypothesis or (2) we fail to reject it.

For example, in the criminal justice system, the null hypothesis is that the defendant is innocent. The prosecution presents evidence, and the jury decides whether the evidence is sufficient to prove the defendant guilty. If the evidence is persuasive, the jury rejects the null hypothesis (defendant is innocent) and accepts the alternative hypothesis: it proclaims the defendant guilty.

What if the evidence is insufficient? The jury does *not* accept the null hypothesis and declare the defendant innocent. Rather, it fails to reject it and declares the defendant not guilty. This distinction between accepting the null hypothesis and failing to reject it is important to remember, because failure to reject the null hyopthesis should always be viewed as a provisional decision: additional evidence may cause us to revise our decision.*

A common approach in statistics is to set a rule for when we would reject the null hypothesis. The actual rule depends on several factors, the least important of which is mathematical. The most important factor is the significance of a wrong decision.

Again, consider our legal case. One error is to declare a person guilty when, in fact, the individual is innocent: a wrongful conviction. In this case, we have rejected the null hypothesis when it is in fact the true state of the world: statisticians call this a *Type I error*, though

---

*In actual legal practice, once a person has been declared not guilty by a jury, no further evidence can be used to change the verdict. Legally, failure to reject the null hypothesis is the same as accepting the null hypothesis.

communications theorists use the more descriptive term *false alarm*. A second type of error is to declare a person not guilty when the individual is, in fact, guilty: a wrongful acquittal. In this case, we have failed to reject the null hypothesis when we should have: statisticians call this a *Type II error* and communications theorists call it a *missed alarm*. The crucial question is, Which is worse?

Society has decided that, for criminal trials, Type I errors are much worse than Type II errors: "It is better that ten guilty persons escape than that one innocent party suffer." Thus, our legal system makes it very difficult for a false alarm (Type I error) to occur; in effect, we must have very persuasive evidence before rejecting the null hypothesis. If this preference applied to the fair coin problem, we might use the following rule: if we see more than 65 heads, or fewer than 35 heads, then reject the null hypothesis and declare the coin to be not fair.

In other situations, we assume the opposite. Fire alarms operate on the principle that false alarms are an inconvenience, while missed alarms are catastrophic. Consequently, we accept that, if we burn toast, the alarm may sound, but most would agree that this is a small price to pay for the assurance that, in a real fire, the alarm will go off. In this case, we make it very difficult for a missed alarm (Type II error) to occur and take even the flimsiest evidence as grounds for rejecting the null hypothesis. In the case of the coins, we might set the rule: if we see more than 51 heads, or fewer than 49 heads, reject the null hypothesis and declare the coin unfair.

The preceding approach tends to work well in the sciences, because we have the luxury of repeating an experiment many times before making a decision. Moreover, science always allows for the possibility that additional evidence may cause us to revise or even reverse our initial conclusion. But, in the real world, we don't always have this luxury. For example, suppose the authorized user has access to sensitive information. The previous approach would require us to collect a certain amount of data before we could make a decision. However, by the time we collected the necessary data, an unauthorized user might have accomplished a great deal of harm. To use our coin problem, we might not be able to flip the coin 100 times, so we'd want some way of deciding whether the coin is fair after just a few flips.

To that end, we can use *Bayesian statistical inference*, which centers on the use of likelihood ratios. One important difference between the two approaches is the following: in the preceding, our two hypotheses were that the coin was fair or that the coin was not fair. In the likelihood ratio approach, our two hypotheses are that the coin is fair or that the coin has a specific type of unfairness.

Suppose we picked our coin from a bag that included some fair coins and some unfair coins weighted so they land heads 60% of the time. Suppose we flip our coin once, and it lands heads. If the coin was unfair, the probability of this occurrence would be 60%; if the coin was fair, the probability of this occurrence would be 50%. The likelihood ratio would be the ratio of these probabilities: 60% to 50%, or 1.2. What this tells us is that our observation is 1.2 times more likely to occur if our coin is unfair.

The significance of the likelihood ratio is most easily understood in terms of the *odds of an event*: this is just the ratio of the probability of the event occurring to the probability of the event not occurring. The likelihood ratio tells us how we should modify the odds based on our observation. If we picked our coin out of a bag with 99 fair coins and 1 unfair coin, the odds of picking an unfair coin are 1 to 99. But if the coin landed heads, giving a likelihood ratio of 1.2, then we should increase the odds by a factor of 1.2, to 1.2 to 99. This isn't a great change. Initially, there was a 1% probability the coin was unfair; with the additional evidence, the probability has increased to about 1.198%.

Suppose we flipped the coin a few more times. If we flipped the coin 5 times and observed it land heads 4 of the 5 times, the probability of an unfair coin doing this is about 26%, while the probability a fair coin does this is 15%. Thus, the likelihood ratio would be 26% to 15%, or 1.66.

Regardless of how many or how few times we've flipped the coin, it's possible to compute the likelihood ratio. If we imagine constantly revising the likelihood ratio as we gather data, then once the likelihood ratio exceeds a certain threshold, we might be willing to conclude that the coin is unfair. As before, we never conclude the coin is fair; that decision is always subject to revision, pending further evidence. This provides an ongoing assessment of whether the coin is fair.

As with standard hypothesis testing, we must consider how sensitive we want our test to be. If we regard locking out an authorized user (or, at the very least, requiring them to resubmit a password) as inconvenient, we might want to set a lower threshold. In the original RAND experiment, one of the typists was locked out because her second digraph timing was not sufficiently similar to her first.

Playing Games

Both of these systems promise to *eventually* lock out unauthorized users who have somehow gained access. However, they don't work well to prevent the unauthorized users from gaining access in the first place. The problem is that trigrams are rare enough that a user would have to type several paragraphs before their identity could be checked; likewise, there are enough variations in where and how people hold their cell phone and how they walk to require extensive data collection before we can use this information to unlock the phone.

One way around this is to require the user to type certain trigrams. We might imagine a person having to type a trigram repeated: *str-str-str-str-str-str-str*. However, there's another problem: if we type a trigram like *str* repeatedly in a short span of time, our speed on the last few repeats is likely to be very different from our speed on the first few and very different from our standard rate of typing the trigram in a text.

A promising alternative was presented in May 2014 by Hristo Bojinov and Dan Boneh, of Stanford University; Patrick Lincoln, of SRI International; and Daniel Sanchez and Paul Reber, of Northwestern University. They described a new system called *Serial Interception Sequence Learning* (SISL), which forces users to prove their identity by playing a game.

During a training session, letters drop from the top of the screen and users must type the letters as they appear (the researchers note the similarity to popular games like *Guitar Hero*). The letters seem to appear at random, but there's a catch: Unknown to the users, the system has selected, completely at random, a 30-character sequence. This sequence will occasionally appear as part of the dropped letters.

For simplicity (and to continue the *Guitar Hero* analogy), imagine that the letters can be one of A through G and that the computer generates a three-letter sequence unique to each user. One user might be assigned the sequence A-B-E. The computer might then present the following letter sequence (a "song" in the game, though in contrast to *Guitar Hero*, the goal is to play the notes as fast as possible):

F-B-F-B-E-E-**A-B-E**-A-B-F-G-B-G-**A-B-E**-
G-B-F-B-**A-B-E**-B-B-B-C-B-B-E-...

Note that the computer-generated sequence appears random but includes the A-B-E pattern that forms the password (which are in boldface to make it easier to see). Because this particular sequence is repeated, users will learn to type it faster than other sequences, like F-B-F.

Now imagine that the user wants to log in. Again, the computer presents them a "song" to play, for example

B-C-G-D-C-A-**A-B-E**-A-D-E-D-C-B-**A-B-E**-

D-G-G-E-B-A-A-E-D-**A-B-E**-D-D-...

The computer then measures how long it takes the user to type the trigram A-B-E. If the user is noticeably faster at typing A-B-E than they are at typing other trigrams, the user is identified as someone who has been trained to type A-B-E: in other words, the authorized user.

There are some interesting features to this system. First, the authorized user doesn't know their password. In particular, they don't know which 30-character sequence gains them entry into the system, and what you don't know, you can't give away, deliberately or accidentally. Second, the system is hardened against dictionary attacks: if the password consisted of 30-note sequences, where each note was one of seven, then the keyspace would have 22,539,340,290,692,258,087,863,249 different passwords. Since the computer, not the user, chooses the password, then every part of the keyspace would be used. Consequently, this type of authentication seems very difficult to break.

The Future of the Password

While the approach of Bojinov and Boneh seems to be a promising way to guarantee the identity of a user, the advantage of a simple password is that it allows quick access to sensitive material. While many users would be happy to play online games as a diversion, it would take significantly more time to log in. Biometric identification, with its requirement of specialized equipment, also seems unlikely to replace the password. And while keyboard forensics and continuous identification are useful for making sure the person accessing sensitive information is the same as the person who logged in originally, they don't exclude unauthorized persons quickly enough. For the foreseeable future, we will need to continue using the password, with all that it entails.

7

\\\\\\\\\\\\\\\\\\\\\\

The Company We Keep

A word should be added about the significance of research of this kind. . . .
It exemplifies a methodological approach which will, we feel, assume a
larger role in the social research of the next decade: namely, making
social relationships and social structures the units of statistical analysis.

JAMES COLEMAN, ELIHU KATZ, AND HERBERT MENZEL
"The Diffusion of an Innovation among Physicians"

In 1953, Lloyd Conover of Pfizer obtained US Patent 2,699,054 for tetracycline, a new, broad-spectrum antibiotic. Eager to expand its business, Pfizer took unprecedented steps, such as taking out multipage advertisements in medical journals. The next year, Pfizer provided a $40,000 research grant to Herbert Menzel, of Columbia University, and James Coleman and Elihu Katz, of the University of Chicago, to study the effectiveness of this approach. The pilot study, of 33 doctors in a New England town, would be expanded to a study of 125 doctors in four Illinois cities.*

Coleman, Katz, and Menzel asked each participating doctor to identify which of their colleagues they turned to for advice, which ones they discussed cases with, and which ones they met socially (the *sociometric questions*). In addition, each doctor was asked which factors

*For reference, see Everett Rogers, *Diffusion of Innovation*, 67.

they considered most important when ranking other doctors, and on the basis of these answers, classified the doctors as profession oriented (characterized primarily by whether they felt the recognition of colleagues was most important) or patient oriented (characterized primarily by whether they felt patient respect was most important). Finally, the prescription records were obtained from local pharmacies to determine when each physician first began prescribing tetracycline to their patients.

Unsurprisingly, the researchers found that the best predictor of when a doctor would prescribe a new antibiotic was the number of prescriptions for antibiotics written by the doctor. Surprisingly, the *second* best predictor of when a doctor would prescribe a new antibiotic was *not* their classification as profession oriented or patient oriented, but rather how they answered the sociometric questions.

For example, participating doctors were asked to identify which of their colleagues they met socially. This allowed the doctors to be classified into cohorts: 33 of the doctors were named as friends by three or more of the others, 56 as friends by one or two others, and 36 had no friends among the participants. Within eight months of the introduction of tetracycline, 90% of the social butterflies had adopted it, in contrast to about 65% of the less social and only 40% of the lone wolves. Similar results held for the other sociometric measures.

Coleman, Katz, and Menzel pointed out another important finding. When comparing the adoption rates of the profession-oriented physicians to that of the patient-oriented physicians, they found that the adoption rate of the former group was *consistently* higher than that of the second group; in fact, the researchers found that in general, when the doctors were separated into cohorts based on *individual* characteristics, there was a consistent difference in the adoption rates. In contrast, when comparing the adoption rates of the highly friended physicians to those friended by fewer or even none of their colleagues, the difference in adoption rate grew, from very little difference at first to a substantial difference within eight months.

What does this imply? First, the consistent difference in adoption rates when individual characteristics are considered suggests that these individual characteristics correspond to an intrinsic inclination to prescribe new drugs. In contrast, the minimal difference *at first* between highly friended doctors and the rest suggests that *at first* these doctors have roughly the same intrinsic inclination to try the new drugs. However, the growth in the difference implies that the highly friended doctors become much more receptive to the new drugs over time, while the rest do not. Put another way: the intrinsic inclination to accept a new drug was important for determining short-term adoption rates, but over time, the extrinsic factors of social integration became more important.

eBay Neckties

The work of Coleman, Katz, and Menzel was an early example of *social network analysis* (SNA), which might be loosely defined as the geometry of organizations. It might seem strange to talk about the geometry of organizations, but we regularly talk about love triangles within our circle of friends, so our language implicitly recognizes such a geometry. Those who study SNA hope to use information about the *network* (the so-called *metadata)* to discern information about its components (in most cases, the people comprising the network). But how?

One such approach appears in US Patent 8,473,422, assigned to eBay on June 25, 2013. To illustrate the method, computer scientists Zeqian Shen and Neelakantan Sundaresan used 78 million transactions between 14.5 million users of eBay to construct a *social commerce network.* As with Google's PageRank algorithm, Shen and Sundaresan analyzed the digraph of buyers and sellers on eBay. In their network, buyers and sellers are represented as nodes, with links connecting sellers to buyers.

Suppose Alice (A) sells something to Bob (B); Bob sells something to Carol (C); Carol sells items to Dan (D); and Dan sells items to both Bob and Ellen (E) (see Figure 7.1). Bob, Carol, and Dan are said to be *connected*: we can follow a link of transactions from any one of them to any other. Alice and Ellen, on the other hand, are not connected: while there's a link *from* Alice *to* Ellen, there's no return path. In the parlance of the eBay patent, Bob, Carol, and Dan are in the *Strongly Connected Center* (SCC): the largest set of connected users. Alice is part of the *In component*: users who link to the SCC. Meanwhile, Ellen is part of the *Out component*: users with links from the SCC.

Nodes might also connect to the In and Out components without being connected to the SCC: thus, Alice might also sell to Frank (F), and Georgia (G) might sell to Ellen. These users form *tendrils.* Finally, it's possible for there to be a path from the In component to the Out

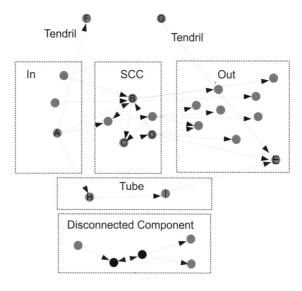

Figure 7.1. Necktie Graph

component that doesn't go through the SCC: Alice might sell something to Harry (H), who sells something to Ingrid (I), who sells something to Ellen. This set of users forms a *tube*. Finally, since the SCC is the *largest* interconnected set, and the In and Out nodes are those that connect to the SCC, it's possible that there are other collections of nodes that are not connected to the SCC: these nodes form disconnected components. The In, SCC, and Out components form a shape evocative of a necktie, giving rise to the term *necktie structure*.*

Overall, 5.83% of eBay users were in the SCC, 3.03% were in the In component, and fully 65.8% were in the Out component. As expected, members of the In component sold the most, and members of the Out component bought the most. However, on a per-user basis, the SCC was the most active: they sold an average of 55.83 items (almost as often as the average of 63.84 sales by In component members) and bought an average of 14.16 items (considerably more than the average of 6.48 purchases by Out component members).

These figures are for the 78 million transactions analyzed by Shen and Sundaresan. But since eBay categorizes sales by type, it's possible to

---

*This is the term used by Shen and Sundaresan, but the term *bowtie* structure is somewhat more commonly used.

analyze the social commerce network of each category, possibly revealing some additional information about consumer purchasing habits. To characterize the social commerce network, Shen and Sundaresan suggest evaluating the *social strength*, a weighted sum of the sizes of the SCC, In, Out, Tendrils, and Tubes:

$$\text{Social Strength} = a1 \times \text{SCC} + a2 \times \text{In} + a3 \times \text{Out} + a4 \times \text{Tube} + a5 \times \text{Tendril}$$

where the values a_1 through a_5 are referred to as *parameters* of the formula. As an example, they suggest using $a_1 = 1$, $a_2 = 0.5$, $a_3 = 0.5$, $a_4 = 0.5$, and $a_5 = 0.25$, which makes the SCC the most important component of the network and the tendrils the least important.

Of what use is such an analysis? The In and Out components correspond to the sellers and buyers in a traditional retail network: they are the stores, and the shoppers at the stores. However, members of the SCC were nearly as active sellers as members of the In component, and far more active buyers than members of the Out component. Since eBay and other auction sites make their money on sales commissions, this suggests that a profitable strategy might be to try and shift as many users as possible into the SCC.

Churning

To be sure, a person's membership in the various components of the social commerce network is likely the result of intrinsic factors that are very difficult to alter. However, there are other uses for social network analysis. One possibility concerns *customer churn.*

Consumer products fall into two broad categories: goods, which are physical items, and services, which are not. Many goods are purchased infrequently: you don't buy a car every month. Moreover, different manufacturers of the same type of good are able to distinguish themselves from each other: some car companies emphasize power and handling, others comfort and luxury, still others fuel economy and reliability. Services, on the other hand, are much harder to differentiate: past a certain level of coverage, one mobile phone provider is like any other. Consequently, the telecommunication industry is plagued by the problem of churning, which occurs when a consumer switches from one provider to another.

One way to limit churn is to provide superior service at competitive rates. Another is to lock users into multiyear contracts that impose stiff penalties for early cancellation. Even so, the churn rate among mobile phone providers is typically between 1% and 5% *per month*, so that over the course of a year, a company could lose more than half its customers! It's clear that additional steps need to be taken to reduce customer churn.

Besides improving service and lowering rates, a company might offer incentives to existing customers. Thus, in exchange for renewing a multiyear contract, you get a brand new phone. But from the service provider's point of view, offering such incentives to loyal customers is a waste of money. Thus a key problem is identifying customers likely to churn and offer the best (and costliest) incentives to just these customers.

What are the characteristics of customers likely to churn? In 2011, Ken Kwong-Kay Wong, now at Ryerson University in Toronto, Canada, analyzed the churning behavior of more than 11,000 customers of a Canadian wireless company. About 50% of these customers switched carriers over a 44-month period. Wong identified *rate plan suitability* as a key reason why customers churn.

Typically, customers pay for a fixed amount of minutes and data. *Breakage* occurs when the customer uses a different amount, and in both cases the company benefits: from overage fees charged to those who exceed their allotment, and from the extra revenue earned by selling a customer a more expensive plan than he or she actually needs.

Wong concludes that companies interested in limiting churn would do well to try and match their customers with appropriate rate plans. Thus, a phone company might analyze a customer's month-to-month usage, and if a customer is constantly paying overage fees, or consistently using far less than his or her allotted data and minutes, the company might offer a more suitable plan in an effort to discourage the customer from switching.

Unfortunately, Wong notes, the *company* has no incentive to do so. His analysis suggests that the Customer Lifetime Value (CLTV) of those on non-optimal plans is nearly twice as great as those on optimal plans. In other words, while offering customers a more appropriate plan could reduce churn, it would also reduce revenue. Customer churn remains a problem, but rate plan optimization is not a profitable solution.

Patently Mathematical

Friends and Family

Thus, rather than trying to retain customers likely to churn, companies instead try to retain customers likely to influence others to churn. This approach is part of US Patent 8,249,231, assigned to IBM on August 21, 2012.

The developers, Koustav Dasgupta, Rahul Singh, Balaji Viswanathan, Dipanjan Chakraborty, Sougata Mukherjea, and Amit Nanavati, of IBM's India Research Laboratory, and Anuma Joshi, of the University of Maryland, used the call data records for a telecomm provider for the month of March 2007. This information allowed them to construct a graph whose nodes were customers of the provider, and where edges existed between customers who made *reciprocal* calls. The focus on reciprocal calls helped eliminate one-way calls (for example, calling a bank to verify a deposit, or a doctor's office calling to confirm an appointment) and avoids the problem of who called whom. Dasgupta's group produced a graph with 3.1 million nodes and 12.6 million edges. Their results, presented in Nantes, France, in 2008, revealed an important feature: you are more likely to churn if the people you call have churned.

The group turned this observation into a discrete time model for identifying subscribers likely to churn. Like Google PageRank, we begin by assuming each node begins with a certain amount of customer dissatisfaction, and that this dissatisfaction is redistributed in some fashion at the end of each time segment. In contrast to PageRank, we assume that when the dissatisfaction exceeds a certain threshold, that customer leaves the network.

From the company's perspective, it's best to intervene *just* before this happens, so the model could be used allow them to identify which nodes are about to churn. For example, we might use the model to identify which nodes have high levels of customer dissatisfaction, so that even a modest increase would cause churning.

It remains to find the *parameters* of the model. Dasgupta's group based their model on spreading activation (SPA) techniques, where the redistribution of the quantity is based on two factors. First, the fraction of dissatisfaction retained by each node (since the remaining dissatisfaction will be distributed to the node's neighbors). Dasgupta's group found that the best predictions occurred with a spreading factor of 72%: in other words, each node distributes 72% of its energy, and retains the remaining 28%.

The second factor is how this dissatisfaction spreads through the node's neighbors. In a simple model, we might allocate the energy equally to all contacts. However, it might make more sense to *weight* the edges, either by call volume or by call time. For example, suppose that X spent 20 minutes conversing with Y, 6 minutes conversing with W, and 10 minutes conversing with Z; likewise, W spent 3 minutes conversing with Y and 6 minutes conversing with Z. We can use this call time to produce an *edge weighted graph* and then use the weights to apportion the energy distribution (Figure 7.2).

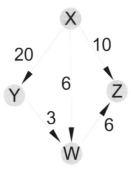

Figure 7.2. Edge Weighted Graph

Suppose X begins with 100 units of energy and all the rest have 0 units of energy. We might describe the energy distribution as a vector: [100, 0, 0, 0], where the vector components give the amounts of energy in X, Y, Z, and W, respectively.

Now we ring the bell, and let this energy move. Since X began with 100 units of energy, and we're using a spreading factor of 72%, then X will distribute 72 units of energy to connected nodes. X communicated for a total of 20 + 6 + 10 = 36 minutes. Of this time, $\frac{20}{36}$ was with Y, $\frac{6}{36}$ was with W, and $\frac{10}{36}$ was with Z. Thus, X keeps 28 units and distributes 72 units; of this amount, Y would receive $\frac{20}{36}$, or 40 units; W would receive $\frac{6}{36}$, or 12; and Z would receive $\frac{10}{36}$, or 20 units. Ordinarily, Y, Z, and W would also distribute their energy to their neighbors, but in our scenario, they don't have any yet. Thus, after one day, the new distribution of energy is [28, 40, 12, 20].

On day 2, each node distributes 72% of its energy and keeps 28%. Thus, Y, which has 40 units of energy, will distribute 72% of 40 units, or 28.2 units, keeping 11.8 units for itself. Since Y communicated with X for 20 minutes and with W for 3 minutes, X will receive $\frac{20}{23}$ of the distributed energy, and Y will receive $\frac{3}{23}$ of the distributed energies. X and W are also receiving energy from Z and from each other, and with only a little effort, we can determine that, at the end of the day, the new distribution of energy will be approximately [45.34, 24.13, 14.66, 15.88]. If we repeat this process, we find that after a very short time, the energy distributions converge to about [40, 25.56, 17.78, 16.67].

There are other ways to distribute the energy. In 2013, Chitra Phadke, Huseyin Uzunalioglu, Veena Mendiratta, Dan Kushnir, and Derek Doran of Alcatel-Lucent developed a more sophisticated model of influence spreading based on *social tie strength*. Their approach would become US Patent 8,804,929, assigned on August 12, 2014, to Alcatel-Lucent.

The intuitive basis behind the approach of Phadke and her team is easy to understand. You're more likely to listen to your friends and family than to an advertisement. If we can identify which nodes represent friends and family, we can make influence from those nodes more important.

But how can we find this information? One possibility is to use information about the users: where they live, work, and play. However, using this information is problematic. First, while the phone company presumably has a billing address and might have a work address, it would not in general have other information about a customer; and this information would be completely unavailable for customers of other companies. Moreover, such information is protected by privacy considerations, so unless the user volunteered this information, the phone company would have legal obstacles to obtaining it.

In contrast, the metadata has fewer protections: A person talking on a cell phone has a reasonable expectation of privacy for the content of the conversation, but the existence of the conversation is considered public information. And by its very nature, the customer voluntarily gives the phone company several crucial pieces of information: the number he or she is calling, the duration of the call, the location from which the call is made, and so on.

This allowed Phadke and her team to determine social tie strength as follows. First, they computed a weighted sum of several factors that presumably contribute to social tie strength. One obvious factor is the total duration of calls during a billing period. But it's possible to spend a great deal of time talking to someone you have no social connection with (for example, your bank). Likewise, it's possible that you *don't* spend a lot of time talking with your friends on the phone, instead limiting your calls to brief conversations like, "Want to meet after work and go to dinner?" To that end, the Bell Labs researchers also considered *neighborhood overlap* of a user's calling circle: you and your friends are likely to call the same group of people, while you and your bank are not. Next, they computed the social tie strength using a formula of the form $w = 1 - a^{-x}$, where x is the weighted sum of the factors.

This model requires choosing several parameters, notably the weights of the different factors that constitute x and the value of a. To determine these parameters, Phadke and colleagues proceeded as follows. Suppose our model calculates social tie strength on the basis of a weighted sum of the call duration and neighborhood overlap. Then we need three parameters: c, the weight for call duration; n, the weight for neighborhood overlap; and a, the value used by the formula for computing social tie strength from x.

We can begin by picking random values for these quantities. Since we want neighborhood overlap to be more important than call duration, we might take $c = 0.1$, $n = 4$, $a = 2$. Now consider the call data records for the IBM approach (Figure 7.2). We calculate the following social tie strengths:

- The link joining X and Y: The call duration is 20 minutes. Note that X also contacts Z and W, while Y only contacts W, so 50% of those contacted by either X or Y are contacted by both, making their neighborhood overlap 0.5. This gives $x = 0.1\,(20) + 4\,(0.50) = 4$, and $w = 1 - 2^{-4} = 0.9375$.
- The link joining X and Z: The call duration is 10 minutes, and 50% of those contacted by either X or Z are contacted by both, so their neighborhood overlap is 0.5. This gives $x = 0.1\,(10) + 4\,(0.50) = 3$, and $w = 1 - 2^{-3} = 0.875$.
- The link joining X and W: The call duration is 6 minutes and there is 100% overlap in who they call, so $x = 0.1\,(6) + 4\,(1) = 4.6$, and $w = 1 - 2^{-4.6} \approx 0.9588$.
- The link joining Y and W: The call duration is 3 minutes, and there is 50% overlap in who they call, so $x = 0.1\,(3) + 4\,(0.50) = 2.3$, and $w = 1 - 2^{-2.3} \approx 0.7969$.

- The link joining Z and W: The call duration is 6 minutes, and there is 50% overlap, so $x = 0.1\,(6) + 4\,(0.5) = 2.6$, and $w = 1 - 2^{-2.6}$ ≈ 0.8351.

This gives us the edge weights in Figure 7.3. We can then apply our SPA model, this time using the social tie strengths to weight the edges, and try to predict which nodes will churn.

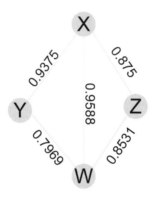

Figure 7.3. Weighted Digraph (Alcatel)

Because we picked c, n, and a at random, it's highly unlikely that our model will predict anything useful. For most of history, this would have been a problem, and much time would have been spent trying to determine c, n, and a from a priori considerations. But in the last years of the twentieth century, high-speed computers made a new approach possible: If the chosen values for c, n, and a gave us a model with poor predictive ability, we could choose a different set of values and produce a different model. We could keep changing the parameters until we found a model that gave us satisfactory predictive abilities. We'll explore this approach in more detail in the next chapter.

By using a data set provided by an (unnamed) large telecomm service provider, Phadke and her team found a model (whose details are proprietary, and so aren't revealed in the patent or in the publications) that seemed to be a good way to identify churners, finding as many as 3.3 times as many churners as other methods.

Eigenvector Centrality

The approaches of Dasgupta's and Phadke's groups can be used to identify customers who are likely to churn. But is this enough? IBM allowed its patent to lapse in 2016, so it appears the approach proved unmarketable. However, with annual churn rates as high as 50%, some approach seems desirable.

One problem is that with such high churn rates, contacting all customers likely to churn would overwhelm any customer retention program. Companies might well be willing to let some customers go, as long as their departure didn't cause a host of other defections. What is needed is some way of identifying critical churners: those whose departure would strongly influence others to leave. The *key player problem* seeks to find the most important nodes in a network.

One possibility is by finding the *principal eigenvector* and its associated *eigenvalue*. The prefix *eigen* comes from the German word *characteristic*, and if we continue to view the network as representation of how some substance can flow through a system, the eigenvectors and eigenvalues (there are usually several pairs) tell us some key characteristics of the system. Thus, Google's PageRank algorithm is based on finding the steady state vector, which is an eigenvector with eigenvalue 1.

Actually, the use of eigenvectors to rank items in a list predates the work of Brin and Page by several years. In 1993, mathematician James P. Keener suggested a method of ranking football teams based on the following idea: The rank of a football team should be a linear combination of the ranks of the teams it plays against and its win/loss record against those teams. Finding the ranks corresponds to finding the principal eigenvectors of a system of equations.

In PageRank and SPA modeling, the quantity being distributed is constant. This makes sense in these contexts: with PageRank, the quantity corresponds to the number of web surfers, while with SPA modeling, the quantity is the amount of energy at a node. But it's conceivable that the quantity in question might not be constant, for example, if we could duplicate it without degrading it. Information is this type of quantity: If you have a digital photo album of 100 pictures, you can send every picture to every one of your friends. It seems reasonable to suppose that customer dissatisfaction has a similar behavior: a truly annoyed customer could post a complaint on social media and have it received, without dilution, by all of his or her associates.

Suppose that in the graph of Figure 7.3, X begins with 100 units of a quantity (we'll call it *gossip*), and that each stage *every* node distributes this total quantity to every other node, regardless of the edge weight. Then our initial gossip distribution vector would be [100, 0, 0, 0]. X would then distribute 100 units to Y, Z, and W (but receive nothing in return, as they have nothing to give back), and our next gossip distribution vector would be [0, 100, 100, 100].*

On the next day, X would have nothing to give. However, it would receive 100 units of gossip from each of Y, Z, and W, giving it $100 + 100 + 100 = 300$ units. We'll ignore the fact that these are the same 100 units of gossip X gave to Y, Z, and W the day before, in part because in a more realistic graph, this "return gossip" forms only a small part of what X would receive.† Meanwhile, Y would receive 100 units of gossip from W, Z would receive 100 units of gossip from W, and W would receive 100 units of gossip from *both Y and Z*. The new gossip vector would be [300, 100, 100, 200].

As before, we can continue trading gossip. Since gossip isn't conserved (alas!), our gossip vector will continue to grow. However, since we're only interested in the *flow* of gossip, and not the total amount, we can normalize the gossip vector. For convenience, we'll divide by the largest component (the so-called *infinity norm*). Instead of starting with the gossip vector [300, 100, 100, 200], we can divide every component by 300 to obtain [1, 0.3333, 0.3333, 0.6667] after rounding.

Continuing in this fashion, we find that after awhile, our scaled gossip vector will converge to [1, 0.7803, 0.7803, 1]. This appears to be a steady state vector. However, there's a subtle difference: If we begin with the gossip distribution [1, 0.7803, 0.7803, 1] and distribute the gossip as described earlier, we obtain a new distribution [2.5630, 2, 2, 2.5630]. This is *not* the same as what we started with. It's only after we divide by the largest component (2.5630) that we obtain "the same"

*We make the simplification that X no longer has the gossip. This is realistic: if X communicates gossip to Y, Z, and W, then X can't communicate the *same* gossip a second time.

†Alternatively, as the game Telephone illustrates, stories change as they move around, so the gossip X receives from Y, which might be based on the same information X initially conveyed to Y, will be different.

distribution. Put another way: If we start with the distribution [1, 0.7803, 0.7803, 1], we'll end up with a new distribution where every component is 2.56 times as great. We say that [1, 0.7803, 0.7803, 1] is the principal eigenvector corresponding to eigenvalue 2.56.

In 2004, Geoffrey Canright and Kenth Engø-Monsen, of Telenor Research Group (a Norwegian telecommunications company), suggested that the eigenvector produces a natural partition of the graph into regions as follows. In PageRank and in similar algorithms, the components of the eigenvector correspond to rankings of the corresponding node, with the larger components having higher rank. If we view these as actual heights, then the eigenvector gives us an "elevation map" of the graph. Thus, we can label each node of the graph with the value of the corresponding component of the principal eigenvector.

In Figure 7.4, each point is labeled with the value of the corresponding component of the principal eigenvector. If each point begins with the indicated amount of a nonconserved quantity (which we'll call gossip), and distributes this amount to *all* of its neighbors, then after all the disbursements have been made, the amounts at each point at the end will be a multiple of their original amount. Thus, the point labeled 0.842 would receive 0.665 + 0.665 + 1 = 2.330 units of gossip, which is 2.769 times as much as it had originally; in fact, all points would have 2.769 times as much as they had originally.

If we view the components of the eigenvectors as actual elevations, we can partition the graph into "mountains" and "valleys" as follows. Consider a point like that with the value 0.652. If we go to any of the connected nodes, we will go to a point with a lower value. In some sense, this point is like a mountain peak: from this point, you can only go down.

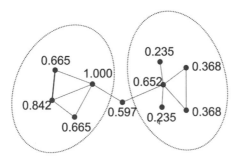

Figure 7.4. Components of Principal Eigenvectors

On the other hand, consider a point like that with value 0.597. If we go to any of the connected nodes, we'll go to a point with a higher value. In some sense, this point is like a valley: from this point, you can only go up.

Now consider something that will spread through the network along the edges. There are two possibilities. First, we might want some things to spread as quickly as possible, such as a new health practice. Alternatively, we might want to limit the spread of a computer virus. In both cases, the *spreading problem* seeks to identify the most important nodes.

Canright and Engø-Monsen suggest that if we want to spread information, we should concentrate on the peaks, which will quickly distribute information to their respective regions. On the other hand, if our goal is to limit some contagion, we might want to focus our protective efforts on the valleys, since an infection from a valley will quickly spread to nearby peaks and from there to the associated regions.

How to Influence Friends

Canright and Engø-Monsen's work forms the basis for US Patent 7,610,367, granted on October 27, 2009, to Telenor. They suggest that identifying the peaks can be useful in helping to diffuse innovation. Typically, this diffusion begins slowly, as new products are discovered by a small group of early adopters; however, eventually the product is adopted by someone with great influence in the social network — a *tastemaker*, to use the marketing term — and thereafter the product spreads rapidly. However, the peaks correspond to those capable of rapid information, so it might be more efficient to target the peaks directly.

We can imagine a similar application in the churning problem. Interestingly, we can approach the churning problem from either direction: We can treat churn as an infection, and try to limit its spread. Or we can treat customer loyalty as an idea we wish to promote. This suggests that customer retention programs should focus on the valleys and peaks, and do their best to retain these customers.

There's just one problem with the approach of Canright and Engø-Monsen: finding the eigenvector for a large network is computationally intensive. Search engine companies can do it, but that's because it's what they're set up to do. In contrast, the resources available to a customer retention department of a telecommunications company are considerably less than those available to Google. This is the fundamental conundrum of social network analysis. In general, determining most network characteristics requires an enormous amount of computational power. Therefore, we want to find a simple-to-compute network characteristic.

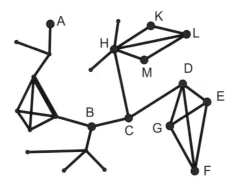

Figure 7.5. Graph

One possibility is the *degree* of a node, which is the number of edges that lead to it. In Figure 7.5, C has degree 3, while A has degree 1. Intuitively, nodes with higher degree are more influential, so if you have a hundred friends, you're more influential than someone who has just five.

However, it should be clear that degree only tells part of the story. In Figure 7.5, D has degree 4 and C has degree 3, but we might hesitate to say that D is more influential than C. This is because, while D is connected to four other nodes, three of those nodes (E, F, and G) are connected only to one another. In contrast, the nodes C is connected to are in turn linked to other sets.

To better determine the influence of a node, IBM researchers Gary Anderson, Mark Ramsey, and David Selby used the idea of *influence weight* in US Patent 8,843,431, granted on September 23, 2014, to IBM. As with Dasgupta's work, we weight the edges to reflect the amount of contact between two people, but this time, to retain information about the directionality of the calls, we'll use a *weighted digraph*, where edges (now drawn as arrows) indicate who called whom, and the weights are the number of calls from one party to the other. Suppose Figure 7.2

shows the number of calls from one party to the other, so that X called Y 20 times, while Y did not call X at all. While in the general case there might be an arrow from Y to X, for illustrative purposes we'll assume that there are no such return calls.

Suppose we want to determine how much influence X has. Some of that influence will be exerted directly, through those contacted by X: in this case, Y, Z, and W. However, there are also indirect influences: If you want to convince a third friend to go to a movie, then it helps if you can convince two other friends. The three of you acting together can be more persuasive than any one of you.

Consider the edge weighted digraph (see Figure 7.2). To calculate the influence weight of X, we make three observations: First, X has made a total of 20 + 10 + 6 = 36 direct calls. Second, consider those called by X: Y has made 3 calls to W, W has made 6 calls to Z, and Z has not made any calls. If X has exerted influence on any of his or her callees, then some of that influence can be transmitted down the line to these secondary callees, and if we are using the number of calls as a measure of influence, then these 3 + 6 = 9 calls should be counted toward that influence. Third, we note that the secondary callees have made calls as well: W has made 6 calls to Z.

In a larger graph, it's possible for this call chain to continue. However, it's reasonable to suppose that the influence of X decreases as the call chain lengthens, and the IBM patent assumes that it's only necessary to extend the chain three steps. Thus, X's influence is determined by the 20 + 10 + 6 = 36 calls to the primary callees Y, W, and Z; the 3 + 6 = 9 calls to secondary callees W and Z; and the 6 calls to tertiary callee Z. Note that calls to Z are counted several times: this reflects the fact that X not only can exert direct influence on Z from the 10 direct calls but also can influence Z indirectly, through intermediaries Y and W.

To calculate the influence weight, the IBM patent values primary calls at 1, secondary calls at 0.3, and tertiary calls at 0.1. Thus, X's influence weight will be $36 \times 1 + 9 \times 0.3 + 6 \times 0.1 = 39.3$. By a similar computation, we find the influence weight of Y will be $3 \times 1 + 6 \times 0.3 + 0 \times 0.1\ 4.8$; the influence weight of Z will be 0 (they called no one), and the influence weight of W would be 6. Thus, X is the most influential.

Several features of this approach are worth noting. In an SPA model, Z would receive a great deal of energy, as the recipient of energies from X and W (and meanwhile, because of the directed graph, having no way to disburse such energies): thus, Z would quickly reach the churning threshold. As far as a telecommunications company is concerned, Z has no influence, so it makes no difference whether Z leaves. If Z leaves because they are saddled with an unsuitable rate plan, the company may have no interest in improving Z's condition, and corporate strategy may even be built on acquiring customers like Z. The short-term profit gained from an unsatisfied customer outweighs the long-term income from a satisfied customer.

On the other hand, consider X. The high influence weight of X means that this customer's level of satisfaction is very important, and even if this customer has only a slight inclination to churn, the company might do well to offer further incentives to X, in the hopes that a disinclination to churn will be communicated to others.

The Power of Word of Mouth

The use of SNA is based on the idea that people talk to each other, and what they say about a product or service makes a difference to potential customers. Advertisers have long been aware that word-of-mouth (WOM) recommendations are more relevant to consumer purchasing decisions than almost anything else. A single consumer, pleased with a purchase and willing to recommend it to his or her friends and neighbors, can do more to sell the product than a multimillion-dollar advertising campaign. The rise and fall of the American car industry is a testament to the power of WOM.

Traditionally, companies focus on the *customer lifetime value* or *customer intrinsic value* (CIV): How much will a customer purchase from a company over their lifetime? If the value exceeds the cost of advertising to the customer, then that customer will be targeted in the hopes that they will buy — and continue to buy — the product. By this measure, advertising money spent on non-customers is wasted.

Or is it? Imagine a relatively common situation: You need to rent a car. There's a good chance that the make and model of the car will be different from what you ordinarily drive, and even the best experience with a rental probably won't cause you to rush out and buy a new car. However, you are now in a position to recommend the vehicle to friends and family who might be in the market. Since this type of advertising costs the company nothing, it should be considered part of your value to the company: it is your *customer network value* (CNV). This is the basis of US Patent 7,941,339, granted on May 10, 2011, to IBM and developed by Joseluis Iribarren Argaiz. In 2015, IBM sold the patent to Japanese e-commerce site Rakuten.

Roughly speaking, a person's WOM influence depends on two factors. First, the likelihood they are to recommend a product in the first place. Statistical analysis of customer behavior can be used to predict this likelihood.

However, even those who are likely to talk might be ineffective if their social network is limited. If the simple graph depicted in Figure 7.5 represents a network of friends, then regardless of how effusively A recommends a product, the CNV of A is limited by the fact that A is only connected with one other person. C might be more taciturn, but the fact that C is better connected to the network means that any recommendations he or she does make will spread far and wide.

Argaiz's approach focuses on several graph features, which we can compute for any node. First, the out-degree of the node, which we've already introduced as the number of links coming from the node. Second, the degree centrality, which is the fraction of nodes in the network connected to the node being considered.

Other quantities require some explanation. In any system where "the distance between two objects" has meaning, a *geodesic* is the path of shortest distance. In graph theory, distance is usually measured by the numbers of edges that we must traverse to get from one point to another. The distance between H and F is 3: from H, we pass through C, D, and then F. In a real circle, the diameter is the greatest possible distance between points; consequently, in a social circle, we'll make the *diameter* of a graph the length of the longest geodesic.

Finally, consider a pair of nodes in the network, both of which are linked to a third node; for example, nodes B and D are both linked to C. The fraction of these paired nodes, which are *also* linked to each other, is called the *clustering coefficient* of the graph. In a social network, it corresponds to the chance that your friends know one another.

Argaiz's patent purports to use graph characteristics to determine the CNV of the nodes in a network. It does so by computing two values: the *network value strength* and the *network value extent*. However, unlike the patents for dating sites, which give explicit formulas computing compatibility or suitability, the IBM patent merely notes that

This algorithm involves, without limitation, arithmetic or algebraic operations such as summation, subtraction, multiplication, division or powers. The mathematical formula of such algorithm is not unique as it depends on the specific network or customer type of interest in each case.*

The patent includes, as an example, the formula

Degree × Degree Centrality × Diameter × Clustering Coefficient

The IBM patent points to a serious problem with allowing mathe-. matical algorithms to form part of a patentable device: the formula is presented as an example of how a component of CNV *could* be calculated, but the patent can be viewed as covering any computation of CNV based on network characteristics.

The problem is that the formula given appears to be meaningless. In particular, the degree centrality is based on the number of nodes connected to the node examined, which is in turn the same as the degree itself. The first two factors of the product measure the same thing. Meanwhile, the diameter and clustering coefficient are the properties of the graph as a whole, so these factors affect all nodes in the same way, which means they are useless for distinguishing between nodes.[†] However, because it included *some* formula for calculating CNV from network characteristics, IBM laid the groundwork for claiming *all* formulas for calculating CNV from network characteristics. Indeed, the patent concludes with

various changes and alternatives may be proposed by anyone familiar with the disciplines involved. Nevertheless, any such changes and alternatives are supposed to be encompassed within the scope of the appended claims as long as their final purpose remains to measure and assign Network Value Indexes to individuals of a set based on their detected message pass-along activity.[‡]

In other words, suppose another researcher were to find a different formula, based on network characteristics, that accurately determined

---

*US Patent 7,941,339, p. 9.

[†]Thus, when looking for which car has the best gas mileage, it's not useful to know that every car has four tires.

[‡]US Patent 7,941,339, p. 10.

CNV. By a literal reading of the patent claim, the researcher could not profit from his or her work. Instead, they would have to pay royalties to IBM to use their own work. While it's not clear whether the patent holder could enforce such a broad claim, the mere threat of a patent infringement lawsuit is often enough to force a settlement.

Why would the patent office even allow such a broad claim, which any mathematician would instantly recognize potentially as preempting research in an entire industry? One answer is that patent examiners are not mathematicians. To become eligible to advise the US Patent and Trademark Office, one must submit evidence of scientific and technical training. Coursework short of a degree is sufficient; however, mathematics coursework is *expressly excluded* as evidence of scientific and technical training.

Again, a simple remedy against such broad claims would be to require including the actual mathematical formula or algorithm, with evidence that the formula works as claimed. However, companies could object: the actual formulas might be viewed as proprietary information, and putting them in a patent document risks a company's preeminence. But we can draw a lesson from how food products are labeled. In food labeling, the ingredients are listed, but the exact amounts are not.

Likewise, we can present a mathematical formula but omit the parameters: thus, the eBay patent could have given the formula for calculating social strength but omitted the values $a1$ through $a5$ (or given sample values that illustrate the formula, while keeping the values actually used by eBay proprietary information).

Similarly, one can imagine a formula that uses network characteristics, say,

$$a \times \text{Degree} + b \times \text{Diameter}$$

where a and b are parameters. The patent could include the formula and sample parameter values, which are sufficiently good to give reasonable results; the actual implementation could use proprietary values, which give the best results.

This approach protects the value of the innovation, because such a patent would cover all possible values of the parameters. However, entirely different formulas would not be covered by the patent, allowing for innovation by other researchers.

Word of Mouth: Strength from Weakness

If we want to use graph characteristics to identify influential users, we need to study how social networks communicate information. In 1973, Mark Granovetter, of Johns Hopkins University, considered this problem from a graph theoretic standpoint. Granovetter's work focused on how people found new jobs.

Consider the graph in Figure 7.5, and suppose D is looking for a new job. First, we see that D is connected to nodes E, F, G, and C (and so it has degree 4). Consider the set of nodes D, E, F, and G. Since every node in this set is connected to every other node, we say this set of four nodes *4-clique*. Intuitively, any member of a clique is closely bound to the other members. We can formalize this by considering the *strength* of a tie: the strength of a tie between two nodes is the fraction of nodes they are *both* connected to. Thus, the connection between D and E is very strong: E is connected to 2 of the 3 nodes D is connected to. In contrast, the connection between D and C is very weak: they have *no* contacts in common.

What Granovetter found was that it was the *weak* ties that were important: "People receive crucial information from individuals whose very existence they have forgotten."* If we think about the problem, we can see why. Since D is strongly connected to E, F, and G, there's a good chance that none of them will be able to offer D any new information: F might know that E's company is hiring, but D would be able to receive this information from E directly. In contrast, if C knows of an opening at B's company, this is information that would have been unavailable to D.

The relevance of weak ties to advertising was established in 1987, by Jacqueline Johnson Brown of the University of San Diego, and Peter Reingen, of the University of Arizona. Brown went on to write "Social Ties and Word-of-Mouth Referral Behavior," which linked consumer behavior to social network analysis. The paper would win the 1987 Robert Ferber Award for Consumer Research.

Brown and Reingen analyzed how three piano teachers in a large southwestern city acquired students. The teachers were chosen because none of them actively solicited students or engaged in mass-media

*Granovetter, "The Graph of Weak Ties," 1372.

marketing: consequently, the only way they acquired new students was through WOM recommendations.

In an approach rather typical of this type of research, Brown and Reingen began by contacting 67 of the current students of the teachers. These students (or their parents, in the case of minor children) were asked how they first learned of the piano teacher. If they identified another person as the recommender, that person was contacted and asked the same question; this backward trace was extended as far as possible, and only ended when a recommender could not recall the original source of the recommendation or identified the teacher as the source. In this way, the initial group of 67 expanded to include 145 persons, 118 of whom were on paths that ultimately led to one of the three teachers. This preliminary phase gave Brown and Reingen 118 persons to serve as nodes in a social network. In follow-up interviews, they then identified the links between the individuals.

Their work showed that both strong and weak ties play important roles in consumer decisions. In particular, weak ties are more likely to be conduits through which information about a product flows; however, the information received through strong ties is more likely to be decisive.

Brown and Reingen also considered homophily, the similarity between the person seeking a recommendation and the person asked for a recommendation. Unsurprisingly, they found that people sought recommendations from others like them. More surprisingly, homophily played little to no role in how the recommendation was perceived.

Viral Marketing

The work of Brown and Reingen was based on WOM in its most literal sense: people talking to one another. But the growth of social media allows for a new type of WOM — sometimes called eWOM — and offers a new marketing possibility. Suppose that, instead of launching an expensive advertising campaign for a new product, a company persuades a carefully chosen group of people whose recommendations will cause others to adopt the product. Because the adoption of the product mimics the spread of a disease through a population, this approach is known as *viral marketing*.

In social network analysis, the *key player problem* is finding the most "important" node or nodes in a network. The challenge is that "important" means different things in different contexts. Thus, instead, we can define various *centrality measures*, which provide some information about a node's role in a network. We can then choose which centrality measure is most relevant to the task at hand.

One obvious measure is the degree of the node, and we've already introduced degree centrality. In a digraph, we can further distinguish between *out-degree* (the number of edges *from* the node) and *in-degree* (the number of edges *to* the node). Two more centrality measures are known as *farness centrality* and *betweenness centrality*.

The farness of a node is the total distance (number of edges) between that node and every other node in the network. If, as seems reasonable, a person's influence on another diminishes based on how many intermediaries this influence must pass through, then the greater the farness of a node, the less influential the node. Frequently, we calculate farness, but use the reciprocal of farness, called *nearness*.

Farness is easy to calculate using *Dijkstra's algorithm*, invented by computer scientist Edsger Dijkstra in 1956. Consider H, the node in Figure 7.5 with the highest degree (6). H is adjacent to six other nodes, namely, K, L, M, C, and two unlabeled nodes, so its distance to all these nodes is 1. Now consider the nodes adjacent to these nodes. For nodes like L, all adjacent nodes H, K, and M have been counted, so we can ignore these nodes (and L as well). C, on the other hand, is adjacent to two new nodes, B and D. This means that H will be distance $1 + 1 = 2$ to these two nodes. We can continue in this fashion, expanding outward, until we've included every node in our network. If we do so, we'll see that H is adjacent to six nodes, has distance 2 from two nodes, has distance 3 from five nodes, has distance 4 from six nodes, has distance 5 from one node, and has distance 6 from two nodes. So its farness will be $6 \times 1 + 2 \times 2 + 5 \times 3 + 6 \times 4 + 1 \times 5 + 2 \times 6 = 66$. A similar computation for all other nodes allows us to conclude that B has the least farness, at 54, and by this measure, B has the most influence.

Betweenness is more complicated. Consider the geodesic between any two points. Anyone along that geodesic plays an important role in facilitating communication, so someone along many geodesics plays an important role in communication. Again, it seems reasonable that someone who sits astride the communication routes between other members of the network will be more influential, so the greater the betweenness, the greater the influence. In our graph, there are a total of 231 geodesics. Of these, 91 pass through H. In contrast, 149 of the geodesics pass through B. Again, by this measure, B has considerable influence.

How do these factors affect a node's influence? In 2008, Christine Kiss and Martin Bichler, of the Technical University of Munich, simulated how information spread through a network. Imagine we "infect" a network by selecting certain nodes and giving them a piece of information. These nodes can then transmit this information to their neighbors, who become new sources of infection. This allows the information to spread throughout the network.

To model this process, Kiss and Bichler assumed that each node would transmit the information to some of its neighboring nodes. However, as the classic game Telephone illustrates, the trustworthiness of information degrades as it passes through more hands: If your friend saw an alligator in the sewer, you might believe them. But if your friend's brother's coworker's cousin saw an alligator in the sewer, you might doubt the story.

To reflect this, we can incorporate a decay factor. Kiss and Bichler consider two possibilities. First, there is *exponential decay* in which the trustworthiness of the message will be t^x, where x is the number of links from the original source. If $t = 1$, then the message retains its trustworthiness as it is passed from person to person; otherwise, the smaller the value of t, the more rapidly the trustworthiness of the message falls. Thus, if $t = \frac{1}{2}$, you would assign trustworthiness $\left(\frac{1}{2}\right)^1 = \frac{1}{2}$ to your friend's observation of an alligator in the sewer, while you'd assign a trustworthiness $\left(\frac{1}{2}\right)^4 = \frac{1}{2} \times \frac{1}{2} \times \frac{1}{2} \times \frac{1}{2} = \frac{1}{16}$ to your friend's brother's coworker's cousin's observation. The other possibility they considered is a *power law decay*, where the trustworthiness will be $\frac{1}{(1+x)^t}$, where again x is the number of links. Thus, if $t = 2$, the alligator story from your friend would have trustworthiness $\frac{1}{(1+1)^2} = \frac{1}{4}$, while the same story from your friend's brother's coworker's cousin would have trustworthiness $\frac{1}{(1+4)^2} = \frac{1}{25}$. In general, an exponential decay decreases more rapidly than a power law decay.

Ordinarily, we would disregard messages with low trustworthiness. However, it's possible you might hear the same story from several people. If your friend's brother's coworker's cousin tells you about seeing an alligator in a sewer, you might disregard it. But if your sister's husband's father and your coworker's brother corroborate the story, you might begin to accept it. Kiss and Bichler assumed that, if the total trustworthiness of the message passed a threshold, it would be communicated. Finally, to keep messages from being repeated forever, they assumed that after some number of transmissions it would be dropped; in a discrete time model, this corresponds to the message being "old news" and no longer of interest.

Kiss and Bichler then applied their model to a real-world network consisting of 54,839 mobile phone subscribers and considered two scenarios: (1) where the information is assigned to a set of randomly chosen customers and (2) where the information is assigned to a set of customers chosen on the basis of various network characteristics.

In the first scenario, 90 randomly selected customers eventually communicated the information to less than 500 others. At the same time, 90 customers selected on the basis of their farness centrality communicated to more than 3,000. If those 90 customers were selected on the basis of their betweenness centrality, they'd communicate to almost 6,000 others. And if the 90 were chosen on the basis of a more complex centrality measure, designated Sender Rank, they would reach more than 8,000 customers. Surprisingly, if the 90 customers were chosen based on their out-degree, the simplest measure, the message would also reach nearly 8,000 customers.

The use of out-degree centrality is very tempting. Other graph characteristics, like farness or betweenness, require having the structure of the entire graph. In contrast, out-degree centrality can be computed for any node by determining how many it's directly connected to; as such, we can compute it even when we have only a tiny part of the graph.

However, the conclusion of which centrality measure is most relevant to finding the influential nodes is dependent on our assumptions about how information spreads through the network. All we know for certain from the work of Kiss and Bichler is that out-degree centrality is a good way to find the most important nodes in a particular network under a particular set of assumptions.

A different approach was taken by Françoise Soulie-Fogelman, which eventually led to US Patent 8,712,952, assigned to KXEN, a data mining company now owned by SAP, on April 29, 2014. Her approach begins with observations of how an actual product spreads through a network. This information, combined with the node centrality measures and other information about the graph, creates a set of *behavioral centrality measures* (though exactly how these are computed is a proprietary process not described in the patent). A model is then produced that will best predict these behavioral centrality measures. Once found, this model can then be applied to the network of interest in an attempt to find the best targets for a viral marketing scheme.

The Tastemakers

Intuitively, we expect recommendations to make a difference. Brown and Reingen's work suggests that network characteristics are important, and advertisers might be able to identify the tastemakers: those whose recommendations will make or break a product. But do such people actually exist?

In 2008, Lisa House and Joy Mullady, of the University of Florida, and Mark House, of Q-Squared Research, surveyed 22 University of Florida students to evaluate their willingness to try three new (hypothetical) products: a candy bar, a new sandwich at a fast-food restaurant, and a new restaurant. Next, the students were asked which of the other participants they knew. This allowed the researchers to construct a social network for the participants. Finally, for each person the student knew, they were asked for the likelihood of trying the new product *if* that person recommended it.

To evaluate the importance of recommendations, the researchers constructed two models to predict the effect of a recommendation. The first model was based only on the respondent's own characteristics, such as sex, age, and whether they were the primary shopper for the household, and how well they knew the recommender. The second model was based on all these characteristics, as well as additional quantities that reflected the network position of both the respondent and the recommender.

For the candy bar, the recommender's farness proved most relevant, though in a somewhat unexpected fashion: the *greater* the farness of a recommender, the more likely it was to positively affect someone's decision to try the candy bar. Put another way, a recommendation for a candy bar from someone who is, by and large, removed from our social group, will have a greater effect than the same recommendation from someone who is much closer to the group. This parallels Brown and Reingen's result that weak links are often used to provide information about a new product or service.

For the restaurant, the recommender's farness also proved relevant, in the same way as the candy bar. However, the respondent's betweenness was also important: the greater the betweenness, the greater the effect of a recommendation. In effect, those who were links between group members tended to be more responsive to restaurant recommendations. Finally, if the recommender was in the core (roughly speaking, the largest clique in the group), the recommendation was more likely to be effective.

The sandwich showed the most peculiar correlations. First, the network characteristics of the recommender had little to no effect on the impact of a recommendation. Second, both farness and membership in the core of the respondent increased the impact of a recommendation: in other words, recommendations mattered to those most removed from the group, as well as those most embedded in the group. Intriguingly, betweenness of the respondent had a negative effect: the greater the betweenness of a respondent, the more likely a recommendation would *decrease* their willingness to try the sandwich.

House and Mullady's work suggests that network characteristics can be used to identify the tastemakers in a group. At the same time, it suggests that *which* network characteristic depends on what we're trying to sell. Thus, if the product is a novelty, whether it's a new candy bar or piano lessons, then farness counts: we seem more likely to try something new if it's recommended by someone we don't know well. However, dining at a restaurant is more of a social activity, and betweenness counts: we seem more likely to heed the recommendation of someone who mediates between group members. Finally, the peculiar correlations exhibited by a sandwich suggest there is much work to be done before social network analysis becomes an indispensable tool for the marketer. However, the question is not *whether* this will happen—but *when*.

8

\\\\\\\\\\\\\\\\\\\\\\\\

The Best of All Possible Worlds

When people used to visit the LEGO Group, one of the things they were told was that there are 102,981,500 possible ways to combine six eight-stud LEGO bricks of the same colour. But one day the Group was contacted by a professor of mathematics who had calculated that this figure was too low.

LEGO COMPANY PROFILE, 2011

On January 28, 1958, at 1:58 p.m., Godtfred Kirk Christiansen submitted designs for a system of interlocking plastic bricks. Originally described as "toy building bricks," the product soon came to be synonymous with the name of Christiansen's company (founded by his father, Ole Kirk Christiansen): LEGO, a contraction of the Danish words *leg godt,* "play well." Since 1964, all LEGO bricks have been made from acrylonitrile butadiene styrene (ABS) and are manufactured with such a high degree of precision that any of the more than 500 billion LEGO bricks manufactured over the past 50 years can be fit together.

One of the key selling points of LEGO bricks is that a child (or an adult — about 5% of total sales go to Adult Fans of LEGOs, known as AFOLs) can use them to assemble an object and then, after the object is no longer of interest, it can be disassembled and the pieces reassembled to form an entirely different object. One of the marketing campaigns for LEGO included the tagline "A new toy every day." In

1974, the LEGO company asked a mathematician to compute the number of ways LEGO bricks can be put together, and they announced the result in a company newsletter: Six of the basic bricks could be put together in 102,981,500 ways.

~~Combination~~ Permutation Meals

Combinatorics is the branch of mathematics that deals with counting the number of ways something can be done. Remember that by the *fundamental counting principle*, if Task A can be done in *m* ways and Task B can be done in *n* ways, then the two tasks can be done in *m* × *n* ways. For example, a lunch deal might allow you to select any of 7 entrées and any of 15 sides; this means there are 7 × 15 = 105 different lunches you could make.

If the choices for the first task are different from those for the second task, we have a *permutation*. However, if the same set of choices can be used for either the first task or the second task, we have a *combination*. To a mathematician, a lunch deal that allowed you to choose a side and an entrée would be a permutation meal; a true combination meal deal would allow you to choose two entrées or two sides.

Combinations are much harder to analyze than permutations. While every permutation question can be answered by the fundamental counting principle, combinations are so complicated that the entire branch of mathematics is named after them. Fortunately, the brick stacking problem is essentially a permutation question. Even so, it's a challenging one.

A LEGO brick is essentially a hollow plastic box. The top sprouts several projections, known as *studs*; when two bricks are stacked on top of each other, the studs of the lower brick extend into the hollow of the upper brick, locking the two together (there are also *tubes* that help secure the studs). Since the connection is purely mechanical, only friction holds the bricks together, and they can be separated with some effort — enough to allow the bricks to hold together, but not so much that they can't be pulled apart. The result of attaching two or more bricks to each other is called a *build*.

The standard brick (insofar as there is one — there are several thousand different types of bricks, not counting color variations) consists of a block four studs long and two studs wide. Imagine having a single

2 × 4 LEGO brick in front of you. How many ways could you attach a second 2 × 4 LEGO brick? Most people might begin by stacking the second brick atop the first, so the two overlap completely; viewed from above, the build looks like an I. Or you could stack the second brick so it is perpendicular to the first at the center; viewed from above, this build looks like a plus sign (+). There are in fact 46 ways you could stack one LEGO brick atop another.

What if we stacked three bricks? To do this, it's helpful to break the problem into two parts. First, we'll find the total number of ways the bricks can be stacked. Next, we'll eliminate the duplicates. There are 46 ways to stack a second brick on top of a first, and 46 ways to stack a third brick on top of the second: that's 46 × 46 = 2,116 ways to stack three bricks.

However, there's a problem. Consider the 46 ways you could stack one LEGO brick atop another. Not all of these are really different. For example, if you set one brick at right angles to overlap the bricks on the bottom, you'll get an L-shaped build. But you could set the second brick at right angles at the top, getting a ⌐-shaped build. While this counts as a different way of stacking, if you turn the ⌐-shaped build around, you'll get an L-shaped build. There are in fact only two builds that can't be turned into another build: the I configuration and the + configuration. The remaining 46 – 2 = 44 builds can be turned into each other, so these represent only $\frac{44}{2}$ = 22 different builds, for 22 + 2 = 24 two-brick builds all together.

You might wonder why we didn't discount these from the start. The reason is that while some two-brick builds can be literally turned into others, we might not be able to do that if we add a third brick. For example, suppose we wanted to form sequences of letters, drawn from the set o, b, d, n, p, q, u, x. There are eight letters, so there are 8 × 8 = 64 two-letter sequences, 8 × 8 × 8 = 512 three-letter sequences, and so on.

However, notice that some of these letters can be literally turned into others: b can rotate to q, d to q, n to u. If we allow the letter sequence to be turned, then some of these sequences are the same: *box* can be turned into *xoq*. We might ask the much harder question: How many distinct n-letter sequences can we form, if we're allowed to turn the completed sequence?

A standard combinatorial approach is to determine the number of permutations and then determine how many of the permutations are duplicates. Consider the two-letter sequences. Every one of these can be literally turned into another: *ob* becomes *qo*, *np* becomes *du*, and so on. In most cases, we'll get a different sequence. However, there are eight sequences that are the same after rotation: *oo, bq, dp, nu, pd, qb, un, xx*.

What this means is that of the 64 two-letter sequences, 8 are distinct and can't be turned into anything else. This leaves 64 – 8 = 56 sequences, half of which can be turned into the other half. These 56 permutations represent $\frac{56}{2}$ = 28 distinct sequences. It follows that the number of distinct two-letter sequences, when we're allowed to rotate the sequence, will be 28 + 8 = 36.

What about three-letter sequences? Again, we might begin with 8 × 8 × 8 = 512 three-letter sequences. There are 16 three-letter sequences that can't be turned into anything else. All the rest can be. There are 512 – 16 = 496 permutations, representing $\frac{496}{2}$ = 248 different sequences, for a total of 248 + 16 = 264 distinct three-letter sequences, when we're allowed to turn them. Proceeding in this fashion, we can determine the number of *n*-letter sequences for any value of *n*.

So how can we compute the number of LEGO builds? Of the 46 × 46 = 2,116 ways to stack three bricks, 4 configurations *can't* be formed by turning the stack; of the remaining 2,116 – 4 = 2,112, half are duplicates of the other half. There are 2,112 ÷ 2 + 4 = 1,060 ways to stack three bricks. If we continue in this fashion, we'll find there are exactly 102,981,504 ways to stack six bricks.

It's possible that a computational error led LEGO to the number 102,981,500. However, a more fundamental problem is that the number only includes *towers*, where each brick is placed on top of the previous brick: this was acknowledged in the original announcement of the number, but when the number was repeated by others, the limitation was not included. Since LEGO bricks can be placed next to each other, the actual number of arrangements is even greater.

How much greater? In 2005, Søren Eilers and Bergfinnur Durhuus, of the University of Copenhagen in Denmark, and Mikkel Abrahamsen, then a high school student, tackled the problem. Eiler and Abrahamsen independently came to the same value: 915,103,765. The combinatorial problem is extremely complex, due to the necessity of eliminating duplicates and the infeasibility of certain combinations of bricks. Durhuus and Eilers write, "But even though symmetry arguments, storage of subtotals for reuse, and other tricks can be used to prune the search trees substantially, we are still essentially left with the very time-consuming option of going through all possible configurations."*

The LEGO Design Problem

The enormous number of ways LEGO bricks can be put together means that even a very small collection of bricks can produce an astounding number of structures. But a more useful question is how to create a specific structure. A recently released LEGO set uses 5,922 pieces to build a scale model of the Taj Mahal. This leads to the LEGO design problem: *Given a structure we wish to build with LEGO bricks, how can we build it?*

Since LEGO bricks have a very specific size and shape, and there is a smallest LEGO brick (the one-third height 1 × 1 brick), the obvious first step is to take the three-dimensional model and *voxelize* it, creating a three-dimensional grid analogous to the pixelation of an image. Unfortunately, the second step — turning the voxelation into a realizable construction from LEGO bricks — is far more complicated.

Part of the problem is the enormous number of ways of constructing any single structure. For example, consider a wall, eight studs in length, one stud thick, and four bricks tall. If we limit ourselves to full-height bricks with dimensions 1 × 1, 1 × 2, 1 × 3, 1 × 4, 1 × 6, and 1 × 8 (there are no 1 × 5 or 1 × 7 LEGO bricks) and require that all pieces have the same color, then each layer of the wall can be built in 114 ways, and so there are 114 × 114 × 114 × 114 = 168,896,016 ways to build this four layer wall!

What makes LEGO design challenging is that not all of these ways are equally good. Thus, we don't want *any* way to build a structure; we want the *best* way to build a structure. Hence, we might ask the following question: *Of all the ways of building the structure, what is the best way to do so?*

This is an example of what mathematicians call an *optimization problem*. Such problems consist of two parts. First, there is a *constraint* (or a set of constraints): what we are trying to achieve, together with any limits on how we may achieve it. In this case, the constraint is that the assembled LEGO bricks must resemble the structure, and the LEGO bricks must be stacked in such a way to produce a stable structure. All

*Durhuus and Eilers, "On the Entropy of LEGO," 435.

solutions must meet the constraints: if we were interested in building a wall four layers high, then even the best possible three-layer wall should not be considered a solution. Next, we need an *objective function*: this is a way of ranking solutions so that we can identify the best solution.

In 1998, LEGO posed the construction problem to Rebecca Gower, Agnes Heydtmann, and Henrik Petersen, three computer scientists. Since most LEGO bricks only connect to those above and below them, certain structures are unstable. If you want to build a wall four studs wide, then you can use four one-stud bricks. However, if the remaining layers also consist of four one-stud bricks, the resulting wall consists of four adjacent columns and has no lateral stability. Gower, Heydtmann, and Petersen were able to construct a *penalty function* that could be used to evaluate a design, based on the overlap between one brick and those above and below it.

The penalty function allows us to compare two proposed designs and select the better design. With computers growing ever more powerful, we might try a brute force solution: Look at all 168,896,016 designs, calculate the penalty value of each one, and pick out the solution with the best design. Unfortunately, the number of possibilities grows rapidly. If a computer could check a million designs each second, it could produce the best design for a four-layer wall in under three minutes; a five-layer wall would take the same computer five hours, and an eight-layer wall would require a little under a thousand years.

The Disassembly Problem

Twenty years after Gower, Heydtmann, and Petersen did their work, the LEGO design problem remains unsolved. Fortunately, while it's true that computers can do things in a few minutes that would take a human being millions of years, it's also true that human beings can do things in a few minutes that would take a computer millions of years. New LEGO sets are designed by master builders, who display a rare combination of artistic talent and engineering savvy.

This leads to a new problem. Once a master builder constructs a set, it's necessary to produce an instruction manual that allows non–master builders to reproduce the set. At first glance, the process might seem easy: take the model apart, brick by brick; the *deconstruction sequence* can then be reversed to form the building instructions.

The problem is that the deconstruction problem is also extremely hard. Consider a very small LEGO build, with just 25 pieces. We can deconstruct it by selecting a piece and removing it. There are 25 ways to select the first piece to be removed, 24 ways to select the second, and so on. Altogether, there are

$$25 \times 24 \times 23 \times \ldots \times 3 \times 2 \times 1 = 15{,}511{,}210{,}043{,}330{,}985{,}984{,}000{,}000$$

possible deconstruction sequences. A computer that could evaluate a million sets of instructions per second would take 500 billion years to evaluate all the deconstruction sequences.

To be sure, we can simplify the problem. We assumed that we could have chosen any of the 25 pieces as the first to be removed. However, in practice, some of these pieces are "inside" the model and are blocked from removal. But while it's easy for a human to determine whether a piece is blocked, it's not as easy for a computer to do so. In 1998, Surendra Gupta, of Northeastern University in Massachusetts, and Askiner Gungor, of Pamukkale University in Turkey, suggested an approach to the deconstruction problem. Gupta and Gungor's approach relies on an *interference matrix*, which represents how certain components block other components from removal.

In order to produce the interference matrix, we need to choose a *preferred direction of removal*. In some cases, the preferred direction is fixed: for example, we might be building a structure from the ground up, in which case deconstruction can only be accomplished by removing pieces from the top. In other cases, we might be able to remove from any direction: this happens when we have a structure we can access from any side. If component i blocks the removal of component j when we attempt to remove component j in the preferred direction, then the entry in the ith row, jth column is set equal to 1; otherwise, it is set equal to 0.

For example, consider the object shown in Figure 8.1, which consists of eight components A through H. If we imagine the components are like puzzle pieces, then they can be placed by dropping them into place; thus the preferred direction of removal is upwards (away from the page). In this case, *none* of the components interfere with the removal of any other component, and the interference matrix consists of all zeroes: the components can be removed in any order.

Suppose the components can only be moved up or down the page. Consider component B. If we try to move it upward, its path will be

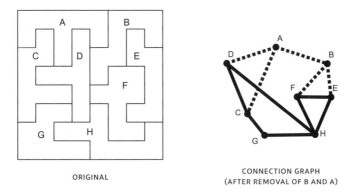

ORIGINAL

CONNECTION GRAPH
(AFTER REMOVAL OF B AND A)

Figure 8.1. Deconstruction Graphs

blocked by component A; if we try to move it downward, its path will be blocked by components D, C, E, F, and H. Because it is blocked by fewer pieces in the upward direction, its preferred removal direction will be upward. If we let the components A through G correspond to the first through eighth rows and columns, then since component A blocks component B in its preferred removal direction, the entry in the second row, first column should be 1; since no other component blocks A's upward movement, all remaining entries in the first column should be 0.

Next, component B (2) has a clear upward path; thus, all entries in the second column of the interference matrix will be 0. We can continue in this fashion to form the complete interference matrix:

	A	B	C	D	E	F	G	H
A	0	1	1	1	0	1	1	1
B	0	0	0	0	1	1	0	1
C	0	0	0	1	0	0	1	1
D	0	0	1	0	0	0	1	1
E	0	0	0	0	0	1	0	1
F	0	0	0	0	1	0	0	1
G	0	0	0	0	0	0	0	1
H	0	0	0	0	0	0	1	0

To use the interference matrix, note that any column of 0s corresponds to a component whose removal is unobstructed: thus, we may easily determine the *removal candidates*. In this case, the first column consists of all 0s, so component 1 (A) is the only removal candidate.

Once we've decided which piece to remove, it no longer obstructs other pieces. Suppose we remove A. Gungor and Gupta suggest the following procedure: First, put a –1 in the row and column entry corresponding to A to indicate that this piece has been removed. Next, since A is no longer part of the structure, then A no longer obstructs any pieces. Consequently, the remaining entries of the *row* corresponding to A can be set to 0. This gives us a new interference matrix:

	A	B	C	D	E	F	G	H
A	-1	0	0	0	0	0	0	0
B	0	0	0	0	1	1	0	1
C	0	0	0	1	0	0	1	1
D	0	0	1	0	0	0	1	1
E	0	0	0	0	0	1	0	1
F	0	0	0	0	1	0	0	1
G	0	0	0	0	0	0	0	1
H	0	0	0	0	0	0	1	0

Now we repeat the procedure. Column 2, corresponding to component B, is the only column of 0s, so it is the only removal candidate. Removing it and modifying our interference matrix gives us

	A	B	C	D	E	F	G	H
A	-1	0	0	0	0	0	0	0
B	0	-1	0	0	0	0	0	0
C	0	0	0	1	0	0	1	1
D	0	0	1	0	0	0	1	1
E	0	0	0	0	0	1	0	1
F	0	0	0	0	1	0	0	1
G	0	0	0	0	0	0	0	1
H	0	0	0	0	0	0	1	0

We might try to complete the disassembly in this fashion. However, this approach leads to a new problem. Consider our 25-piece build. If we were to show the assembly brick by brick, we'd need to show 25 steps. This is feasible for a small build, but most LEGO sets have hundreds of pieces, and many have over a thousand. To avoid unwieldy instruction booklets, each step must show the placement of several pieces. In the real world, large objects are built by constructing *sub-assemblies*, which are then joined together. Jakob Sprogoe Jakobsen,

Jesper Martin Ernstvang, Ole Juul Kristensen, and Jacob Allerelli extended this approach to the LEGO world, which would receive US Patent 8,374,829 on February 12, 2013.

As before, the interference matrix identifies candidates for removal. Since LEGO pieces do not require special tools to attach or remove, nor do specific LEGO pieces require special handling, we can focus on identifying the best subassemblies for removal. To do this, we need information on other structural properties.

The easiest way to do so is to use a *connection graph*, where two nodes are linked if they are physically connected. If the figure shows a LEGO structure, then pieces only link to those above or below them, so our connection graph will show A linked to B, D, and C; B linked to A, E, and F; and so on. (Figure 8.1, Connection Graph) We can represent the removal of a piece by eliminating the link that connects it to other pieces. When we removed components 1 and 2 (A and B), we eliminated all links leading to them (the dashed lines in Figure 8.1, Connection Graph, showing the removed links).

If we examine our connection graph at this point, we see that removing H will split the graph into two parts: the part consisting of F and E, and the part consisting of D, C, and G. In LEGO terms, component H is called an *articulation element*, and these are removed preferentially. The next piece we remove will be H.

Mountain Climbing and Simulated Annealing

While the interference matrix and connection graph help identify pieces for removal, and it's possible to weight different removal strategies, finding the best removal sequence remains a very complicated optimization problem. To help solve the problem, we can introduce a *disassembly graph*.

A disassembly graph can be produced in several ways, though we'll focus on a more intuitive approach using *disassembly states*. Imagine taking a car apart. We can document our disassembly by taking a sequence of pictures. Our initial disassembly state would be a picture of the intact car. If our first step was to remove the hood, our next picture would show the car, without hood, with the hood to one side. Next, we might remove the driver's side door, so the next picture would show the car with the door and hood removed and off to one side.

Note that from any disassembly state, it's possible to make different decisions over what piece or part to remove next. Instead of removing the hood first, we might have removed the driver's side door first. Then we'd have a different disassembly state that shows the car with the driver's side door removed (but the hood still attached). In principle, we'd want to identify all possible disassembly states, though in practice, some will represent impossible or infeasible disassemblies: while you could start the disassembly of a car by removing the tires, it's probably more practical to remove the doors or hood.

We can form a weighted digraph from the disassembly states by joining two disassembly states whenever it's possible, by removing a piece or a part, to go from one to the other, with the weights corresponding to the difficulty of that particular removal. The result will look like Figure 8.2. Here we start with a completed object (A). We can remove a single piece to produce subassembly state B; this requires some effort, so we assign the link a weight of 2. Alternatively, we might start with our intact object and remove a two-piece component to produce subassembly state C; however, this requires more effort, so we assign this link a weight of 5. There may be other disassembly states we're not showing, corresponding to infeasible or impractical disassembly sequences.

From B, we might only have one feasible disassembly step, taking us to G. But if we started our disassembly by removing the two-piece component, we might have several possible steps. We could disassemble the two-piece component, which would also take us to G; we could remove another piece from the main assembly, taking us to D; or we can remove another two-piece component, taking us to E. These removal steps would require different amounts of effort, all reflected in the weighting of the links. If we continue removing parts in this manner, we will eventually arrive at L, where the object has been disassembled into its component parts.

This gives us our optimization problem: Find the path from A to L that has the least total cost. The obvious approach is to try each possible disassembly sequence and select the sequence with the lowest cost. In this particular case, there are seven sequences, and it would be easy enough to check all seven. But as with most combinatorial problems, the number of possibilities grows very rapidly. Thus, we need a more efficient way to solve the optimization problem.

The problem is that if we insist on finding the *best* solution, then

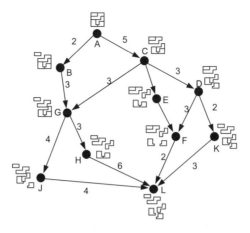

Figure 8.2. Disassembly Graph

we must look at *all* solutions, for how else can we know our solution is the best? In some cases, it might be possible to use the techniques of calculus to find optimal solutions. Unfortunately, most optimization problems are not so neat, and it seems we are condemned to using brute force methods and waiting billions of years for our answers, as long as we insist on finding the best solution.

But in many cases a nearly optimal solution is good enough. If the best possible sequence requires 25 minutes to complete, we might accept a design that requires 27 minutes if the latter can be found in minutes while the former would take a millennium. Alternatively, if we look at other sequences and find they require 30 minutes, 35 minutes, 57 minutes, and so on, then the sequence that takes 27 minutes to complete is the best of the sequences we've looked at. We don't really need the best solution; we need a solution that is good enough. We say we are trying to solve the *approximate optimization problem*.

The approximate optimization problem can be solved in several ways. One of the oldest is known as *mountain climbing*. Imagine that we're trying to maximize our height (mountain climbing), and assume that the only information we have about our height is an altimeter. We can proceed as follows:

1. Choose a random direction.
2. If a step in that direction increases your height, take a step in that direction, and repeat step 1.

3. If a step in that direction reduces your height, go back to step 1 *unless* you've exhausted all possible directions of travel.

The algorithm ends when no direction increases your height; at this point, you can declare you've found a highest point.

In the disassembly graph (see Figure 8.2), we begin at A and can choose to go to either B or C. Going to B has a cost of 2, while going to C has a cost of 5. Since we're actually trying to minimize cost, we choose the direction with the lower cost and head toward B. Once we're at B, there is no choice but to go to G and incur a cost of 3. At G, we can either go to J (cost 4) or H (cost 3), so we go to H. Again, once we're at H we have only one choice and must go to L (cost 6). Thus mountain climbing gives us the disassembly sequence ABGHL, for a cost of 2 + 3 + 3 + 6 = 14.

One advantage to mountain climbing is that we don't have to know what the graph looks like beforehand; we only need to know our choices at each stage. It wasn't necessary for us to know what would have happened if we'd gone toward C instead of B.

However, this is also a disadvantage. Mountain climbing finds *local optimum values*. In real mountain climbing, these would correspond to points higher than anything around them. In some cases, a local maximum is enough. The second tallest peak in a range is almost as high as the tallest. But in others, the local maximum might be nowhere near the *global optimum*. If you end up at the peak of a foothill, you're likely to be much lower than you would be at the peak of the second (or even third) highest mountain in the range.

The problem is that it's sometimes necessary to go down to go up: If you're on a foothill, the mountain climbing algorithm would drive you toward the top of the foothill, but what you really need to do is to descend into the valley before ascending the slopes of the nearby mountain. We say that to find the optimal solution, it's sometimes necessary to make *suboptimal choices*.

When we arrived at G, we had the choice between the path to J (cost 4) and the path to H (cost 3). Selecting the lower cost path compelled us to take the edge from H to L (cost 6). If we'd taken the higher cost initial path, we could complete our trip by taking the edge from J to L (cost 4), lowering our overall cost: ABGJL would cost 2 + 3 + 4 + 4 = 13, instead of 14.

In fact, the least cost path is ACEFL, which requires making two suboptimal choices: toward C instead of B (cost 5 instead of 2) and toward

E instead of D (cost 4 instead of 3). By making these two suboptimal choices, we're able to find a disassembly sequence of cost $5 + 4 + 1 + 2 = 12$.

We can incorporate suboptimal choices using *simulated annealing*, which is modeled on how crystals form as a liquid solidifies. The basic steps are similar to mountain climbing: Wherever you are, pick a random direction and see what happens if you take a step in that direction. If a step in that direction increases your height, take it.

But what if a step in that direction *doesn't* lead to a better solution? In a mountain climbing algorithm, you'd never take a step in a direction that leads to a worse solution. In simulated annealing, you might still take that step. Whether or not we take a seemingly suboptimal step is determined randomly. The net effect is that you will sometimes leave a foothill to search for a higher mountain.

Thus, if we began our disassembly sequence at A, we would again collect the information that the path to B has cost 2, while the path to C has cost 5. In mountain climbing, we'd pick the path to B; in simulated annealing, we have a chance of selecting the suboptimal path to C, and thus some possibility of obtaining the optimal deconstruction sequence ACEFL.

How to Move a Museum

One application of simulated annealing comes from how information travels along the internet. Consider another method of delivering information: the Postal Service. In 1963, the US Postal Service introduced ZIP (Zone Improvement Plan) codes to speed mail delivery. In a five-digit ZIP code such as 02134, the first digit (0) identifies a region of the country; the second two digits (21) identify a subregion; the last two digits (34) identify a specific post office within that subregion. In this way, a letter can be quickly directed to the right post office, at which point the address on the letter can be used to deliver it to a specific location.

We do something very similar for sending information on the internet. Web pages are specified by a URL (*uniform resource locator*): for example, Amazon.com. When you type a URL, a DNS (*domain name server*) translates the URL into an IP address: for example, 46.137.224.14, which identifies a computer just outside of Sydney, Australia. Just like ZIP codes, the individual numbers of the IP address (in this case, 46, 137, 224, and 14) give progressively more detailed information about the identity of the destination computer. For information such as a book order, to get to the computer, it must be sent to the "46" network, then to the "137" subnetwork, then to the "224" portion of this subnetwork, which would then route it to the "14" computer.

However, it's possible that the communication line between computers is down or busy. In that case, the signal would have to wait until the line was repaired or freed up. Alternatively, you could find a different route. For example, suppose you're traveling between Boston and New York, and there's an accident east of New York. You *could* drive back to Boston, then to Chicago, then to Washington, DC, and then to New York, approaching it from the west and avoiding the accident.

In ordinary physical travel, this would be pointless, since the accident would be cleared up long before you completed the alternate route. But signals travel along internet at the speed of light. Going from Boston to New York directly would take 0.001 seconds, while going from Boston to Chicago to Washington, DC, to New York would take 0.01 seconds. Even if the signal had to travel around the world, it would only take a little over 0.1 seconds. Since repairing the line is likely to take minutes to hours, then the circuitous route is going to be the faster one.

We can go further. Imagine we're trying to move something very large: for example, suppose the Museum of Fine Arts in Boston loaned part of its collection to the Metropolitan Museum of Art in New York. The contents might be sent in a single, very large truck. However, if too many cars try to get on the road at the same time, the result is congestion and a traffic slowdown. Moreover, some roads might simply be too narrow for the truck to pass.

The internet faces a similar problem: the amount of data a line can handle is limited. It's as if the road from Boston to New York could only take a certain vehicle weight per hour. Rather than sending our collection in a single large truck, where we'd have to wait until the road was clear, it would be better to divide it among a large number of cars, and send each car out along the first available road. Again, some packages might have a longer distance to travel, but at the speed of light, they will get to the destination without noticeable delay.

We can go further still. Imagine our museum curator in Boston simply putting the items into the first available car going *anywhere*. If the car got to its destination but the package didn't, then the package would be switched to the next available car, again going anywhere. As long as the package kept moving, it would eventually get to their destination. It's as if you organized a house by picking up an item that didn't belong in a room and moved it to the next room: if you kept doing this, then everything would eventually end in the right room.

For example, consider a simple internet, consisting of four computers (*nodes*), which we'll simply identify as 1, 2, 3, and 4. Suppose a message needs to travel from node 1 to node 4. Sophisticated *router software* determines which node the message is sent to, based on available network capacity.

It's possible the message is sent directly to node 4. When node 4 receives the message, its router software verifies that it's the destination of the message and doesn't send it to another node. However, if the message is sent to another node, say, node 3, then the router software first identifies that it's not the destination node and passes it on to another node. Before the message is sent on, it's marked as having passed through node 3; in effect, it's postmarked.

Suppose node 3 passes the message to node 2. First, the router software recognizes that it's not the destination node, so it prepares to send it to another node. But there's no point in sending it to a node the message has already gone through, so the router checks the message history. The message originated in node 1 and passed through node 3, so neither node is an acceptable destination. This means the router at node 2 probably *won't* send the message to nodes 1 or 3. In this case, the only node it can send the message to is node 4.

Again, in the real world, the extra time required for a circuitous route would render this approach to deliver impractical (though the service industry mantra "Don't waste a trip" is a way to increase the efficiency of waitstaff). But with messages traveling at the speed of light, email could travel 10 times around the world before you finish reading this sentence.

This approach to package delivery allows the internet to work quickly and reliably and means that major service outages occur only rarely and are localized; thus, an accident on the road in front of the Metropolitan Museum of Art would prevent delivery of the collection from the Museum of Fine Arts, but accidents farther away would not. However, it leads to another problem.

Suppose we wanted to improve internet service. One way is to improve router software. This is a vast field, and the resulting algorithms are typically based on some method of solving the approximate optimization problem.

The other is to improve the links between the nodes: in effect, build extra roads to improve traffic flow. Consider our simple internet with four nodes. Let $m_{i,j}$ be the number of messages sent from node i to node j; for example, $m_{2,3}$ would be the number of messages sent from node 2 to node 3. The collection of values $m_{i,j}$ forms the *traffic matrix* and gives us a complete picture of how information moves through the network.

How can we find the traffic matrix? In the real world, the nodes might represent different places in a city, and the messages would correspond to the number of drivers going from one location to another. A traffic engineer could construct a mathematical model for traffic flow, taking into account how drivers choose routes. Fortunately, most drivers take the simplest and most direct route, so constructing the mathematical model is easy. The traffic engineer could then alter the parameters of the model, for example, by adding an extra road, and see how the traffic pattern changed.

Now suppose we're trying to model internet traffic. We can do the same thing, but there's a problem: the rules that we would have to incorporate into our simulation are the *same* rules used by the routers. In effect, our simulations of the internet would have to be as complex as the real internet.

There's another problem. This approach requires us to use existing router software to simulate the internet. But router software, like all other software, is constantly improving. Even if we can model today's traffic pattern, there's no guarantee that this will be the same as tomorrow's traffic pattern.

We might learn a lesson from what happened after GPS navigation devices became an essential part of travel. As long as humans were making navigational decisions, they found routes based on simple criteria. But when the problem of navigating was turned over to a GPS system, it found more optimal routes and drivers followed them. Passing through a quiet residential neighborhood might be identified as part of the best route between two cities, causing a significant increase in traffic along the road.

What's important to recognize is that the optimal route exists independent of our ability to find it. Rather than focus on finding the traffic patterns, we might instead look for the optimal routes and assume that these routes will be found somehow; if not by the current generation of GPS or router software, then by some future generation. This is the approach taken in US Patent 7,554,970, assigned to Alcatel-Lucent on June 30, 2009, but subsequently acquired by Nokia when the Finnish telecommunications giant bought Alcatel-Lucent in 2016.

The inventors, mathematician Peter Rabinovitch and computer scientist Brian McBride, describe a method they call Simulated Annealing for Traffic Matrix Estimation (SATME). SATME doesn't actually find *the* traffic matrix for a network; instead, it finds the *optimal* traffic matrix. In effect, rather than considering how information actually flows along a network, they found how information would flow along a network using the best possible routing software. In a similar way, urban planners might expect little traffic along a street and so zone it as a residential neighborhood. But if they determined the street would be identified as part of the optimal route between two popular destinations, they would anticipate heavy traffic along it and might instead zone it for commercial or industrial use.

To find an approximation to the traffic matrix, Rabinovitch and McBride used the following approach. Consider our internet of four nodes. The traffic matrix would consist of the values for the $4 \times 4 = 16$ entries of the traffic matrix.

The actual solution to the optimization problem would be the 16 values $m_{i,j}$. However, in order to apply an optimization method to the problem, we must construct the objective function. In the work of Rabinovitch and McBride, the objective function is a penalty function consisting of three parts, and the goal is to minimize the value of the penalty.

First, as with a real-world traffic problem, we can make assumptions on how many messages originate from a node. Let's designate the number of messages originating at node 1 to be O1 (and similarly for the other nodes). Now consider the actual traffic. Since $m_{i,j}$ is equal to the number of messages sent from node i to node j, and nodes don't send messages to themselves, then the total traffic sent *from* node 1 will be $m_{1,2}$ + $m_{1,3}$ + $m_{1,4}$. Obviously, we want this to be equal to O1. It follows that any difference between this sum and O1 should contribute to the penalty.

Note that it might be possible for the sum to be *greater* than O1. In other words, for us to require more outgoing messages than exist. This may seem problematic; however, a procedure like mountain climbing or simulated annealing can consider impossible solutions as intermediate steps. If you're climbing a mountain in France, it's possible that the easiest path will take you through Switzerland.

While the possibility that one of the provisional solutions may require more messages leaving a node than originate at the node need not concern us, we do have to worry about how such a situation would affect the penalty function. In particular, since we seek the lowest penalty possible, then using $O1 - (m_{1,2} + m_{1,3} + m_{1,4})$ would make it desirable to have the number of outgoing messages be far greater than the number of messages originating at the node, since this would make the difference negative, reducing the penalty. At the same time, using $m_{1,2} + m_{1,3} + m_{1,4} - O1$ would make it desirable to have the number of outgoing messages be zero, since again this would make the difference negative and reduce the penalty. Thus, Rabinovitch and McBride use a standard approach and consider the *square* of the difference. Squaring has two purposes here. First, it eliminates the effect of the sign, since the square of either a positive or a negative number is positive. Second, it increases the penalty for large deviations from the desired value.

In the same way, the penalty function has a second component: the number of messages coming *into* a node. If there are $D1$ messages coming into node 1, then the square of $D1 - (m_{2,i} + m_{3,i} + m_{4,i})$ should be a component of the penalty function.

Finally, a third component of the penalty function should reflect any other constraints. For example, suppose the link between nodes 1 and 2 has a maximum capacity of five messages. Then the penalty function might include these terms $(m_{1,2} + m_{2,1} - 5)^2$.

These components need not count equally toward the total penalty. Rabinovitch and McBride distinguish between *hard constraints*, like those associated with the number of outgoing and incoming messages, and *soft constraints*, like those associated with link capacity. A hard constraint corresponds to a term of the penalty function with a high weight. We might include the term $100 (m_{1,2} + m_{1,3} + m_{1,4} - O1)^2$, multiplying the square of the difference between the provisional outgoing traffic to node 1 and the expected outgoing traffic by 100. Meanwhile, the soft constraints have low weights; we might include the term $10 (m_{1,2} + m_{2,1} - 5)^2$, where the deviation of the traffic from node 1 to node 2 has a much smaller effect on the penalty function. The value of these weights is completely arbitrary and depends on the relative significance of the two types of deviations. As we noted earlier, they also offer a way for a company to continue to benefit from its own innovations: the actual parameter values can be treated as proprietary secrets and need never be revealed, even when the actual formula (without the parameters) is included in publicly available documents.

To run SATME, we begin by assigning randomly chosen values to each of the components $m_{i,j}$. While an optimization algorithm should work no matter where we start, it's helpful to start somewhat close to the solution: You'll get to the top of the Matterhorn faster if you start in Switzerland than if you start in Russia. We'll choose $m_{i,j}$ to be a value less than O_i and D_j: in other words, whatever the initial value of $m_{i,j}$, it will be less than the number of outgoing messages and less than the number of incoming messages. In real terms, the network is operating below capacity. This set of values for the components of the traffic matrix forms our first provisional solution.

Next, we compute the value of the penalty function. Then we take a step by choosing one of the components of the traffic matrix, increasing or decreasing it slightly: thus, we might flip a coin, and if the coin lands "heads," the component will be increased; otherwise, the component will be decreased. This gives us a new provisional solution, and we have to decide whether we should change to the new provisional solution.

How can we make this decision? One possibility is to keep whichever of the two solutions has a lower penalty value. If this is the only rule determining which solution to keep, we have a pure mountain climbing algorithm approach to solving the optimization problem.

However, simulated annealing allows us to occasionally choose suboptimal solutions. Finding the allowable degree of suboptimality is more of an art than a science. Moreover, to carry over the analogy with crystal formation, the allowable degree of suboptimality should decrease as the algorithm proceeds. Thus, we might accept a first step that causes the penalty function to increase by 10%, but we might insist that by the hundredth step, no more than a 1% increase is allowed.

Evolution in Action

Both mountain climbing and simulated annealing are, in some sense, ways to obtain a single strategy. If we use mountain climbing on the preceding, we'll obtain the deconstruction sequence ABGHL, and if we apply it a second time to the same problem, we'd obtain the same sequence.

Simulated annealing offers the possibility of different solutions to the same problem by making suboptimal choices. But in general, simulated annealing relies on making *slightly* suboptimal choices. Thus the optimal path ACEFL would require choosing an initial step of cost 5 over an initial step of cost 2. More likely, a typical simulated annealing strategy would produce some variation on the mountain climbing path ABGHL. When at G, mountain climbing necessarily chooses the path to H (cost 3) instead of the path to J (cost 4); because the path to J is only slightly suboptimal, simulated annealing has a chance of selecting it and producing the lower cost sequence ABGJL.

Another set of optimization algorithms are inspired by biological processes. In general, these optimization methods rely on producing different solutions without regard to whether they are optimal, then combining them in some fashion.

One approach is a *genetic algorithm*, modeled after how species evolve. In biological evolution, each organism has its own genetic code, which allows it to survive in its environment. Because of random variations in the code, some individuals are better fitted to the environment; these organisms are more likely to pass on their genetic code to their offspring.

Genetic algorithms are particularly useful when the objective function itself is hard to describe. US Patent 8,825,168, granted on September 2, 2014, to Cochlear Limited, an Australian biotechnology company specializing in hearing aids, describes how to use a genetic algorithm to fit medical devices to patients.

Because every person is different, many medical devices cannot be mass produced, but must instead be custom designed to fit a particular user. Moreover, while a sufficiently detailed X-ray can be used to reproduce the physical size and shape of a body part, many devices rely on a complex interaction between body and brain. For example, a cochlear implant works by converting sounds into an electrical signal, which is then used to stimulate the auditory nerve, either electrically or mechanically. However, the exact nature of the stimulus cannot be determined from first principles, nor can it be found through a sequence of simple measurements. Instead the inventors, electrical engineers

Sean Lineaweaver and Gregory Wakefield, use a genetic algorithm to fit such devices to patients.

Suppose the cochlear implant stimulates the auditory nerves electrically. There are several parameters that control this stimulation, but for illustrative purposes, we'll focus on two. First, there is the *pulse rate*: how often the nerve is stimulated with an electrical pulse. Next there is the *pulse duration*: how long the nerve is stimulated. Thus, the cochlear implant might stimulate the auditory nerve 1,000 times per second, with each pulse lasting 25 microseconds. Clearly, there is a broad range of possibilities for both. Since the goal is to improve the patient's ability to hear, an audiologist must find the best *parameter map*.

To do so, the audiologist runs the patient through a sequence of tests to determine the THR (threshold, the softest sounds that will be detected by the patient) and MCL (most comfortable loudness, where a sound is loud but comfortable). We might imagine the audiologist setting the pulse rate and duration by turning dials and having the patient judge whether a setting provides better or worse hearing. For example, the audiologist might try the setting of 1,000 pulses per second, with each pulse lasting 25 microseconds, corresponding to setting dials to 1, 0, 0, 0, 2, 5.

The problem is that there are so many possible values for the parameters that it's not feasible to go through all of them. Even if we could, it wouldn't necessarily do any good: the optimal value for THR may cause us to have a very suboptimal MCL and vice versa. It is up to the patient to decide what balance of THR and MCL he or she desires.

After some time, we may end up with a set of pulse rates and duration that give acceptable THR and MCL levels. For example, suppose the patient approves of a pulse rate of 1,000 and pulse duration of 25 microseconds, as well as a pulse rate of 950 and a duration of 38 microseconds, while rejecting a pulse rate of 1,500 and a duration of 30 microseconds. The dial settings correspond to the gene sequence of the parameters: in this case, the 1-0-0-0-2-5 and 0-9-7-5-3-8. Some gene sequences, such as 1-5-0-0-3-0, would be eliminated. In evolutionary terms, these correspond to organisms ill suited to the environment. The acceptable combinations of pulse rate and duration would form the initial population; the unacceptable combinations would be eliminated.

Next, we apply some *genetic operators*. By analogy with natural evolution, there are two. First, there is *mutation*, where one or more genes (digits, in this case) are randomly altered. We might change 1-0-0-0-2-5 to 1-0-1-0-2-5. Alternatively, we can use *crossover*. This involves taking two gene sequences and randomly selecting a cut point, then swapping the portions of the sequence after the cut point. Thus, if we crossed 1-0-0-0-2-5 and 0-9-7-5-3-8, we might swap the last three digits to obtain new sequences 1-0-0-5-3-8 and 0-9-7-0-2-5. This provides us with a new set of pulse frequency and durations to evaluate: in this case, frequency 1010 at 25 milliseconds, frequency 1005 at 38 milliseconds, and frequency 970 at 25 milliseconds.

As in natural selection, there may be *lethal mutations*: for example, 970 at 25 milliseconds may have very poor THL and MCL levels and be eliminated. However, some of these may have even better THL and MCL levels than before. By continuously selecting the best sequences and letting evolution run its course for many generations, a genetic algorithm can lead to parameters providing near-optimal THL and MCL levels.

A Bug's Life

Evolutionary algorithms are computer simulations of what nature has done for a billion years: living organisms *are* solutions to optimization problems. It follows that we can find other ways of solving optimization problems by looking at how life has solved those problems.

In 1992, Italian computer scientist Marco Dorigo, then at the Université Libre de Bruxelles, suggested using an *ant colony system* (ACS) as a way to solve optimization problems. ACS is modeled after the method by which ants find food and is, in the most literal sense, a genetic algorithm, based on the end result of a hundred million years of biological evolution.

In general, ants will wander from the colony in search of food. Initially, these wanderings are random. But when an ant finds food, it returns to the colony, laying down a pheromone trail. Because the pheromones evaporate, more recently used trails will have higher pheromone concentrations. Moreover, since shorter trails take less time to travel, pheromones will have less time to evaporate, and so the pheromone concentrations on these trails will be higher still.

How can we apply this to the disassembly sequence? Imagine we have a number of ants (*agents*, in the language of computer scientists). In the setup phase, we'll have each ant crawl along a randomly selected path and determine the total cost along that path. Next, the trail taken by each ant will be painted with a "pheromone trail," whose intensity will be inversely proportional to the cost: the higher the cost, the lower the pheromone levels.

Now we repeat the process, but this time the path an ant selects will be randomly determined based on the pheromone level along that path. As before, paths ants travel along are repainted with a pheromone trail. Moreover, some "evaporation" occurs, and pheromone levels along all trails are reduced by some amount. If we repeat this process many times, the lowest cost trails will have the highest pheromone levels (and, consequently, the highest ant traffic along them).

Consider our disassembly graph (see Figure 8.2), and suppose we have four ants. The first ant travels along the path ABGHL, the second along ABGJL, the third along ACDKL, and the fourth along ACEFL. Each edge along a path is painted with a pheromone level inversely proportional to the cost of the path. Each edge along the path ABGHL, which has total cost 14, will be given a pheromone level of $\frac{1}{14}$; each edge along path ABGJL, which has total cost 13, will be given a pheromone level $\frac{1}{13}$; each edge along path ACDKL, which has total cost 13, will be given a pheromone level of $\frac{1}{13}$; and each edge along path ACEFL, which has total cost 12, will be given a pheromone level of $\frac{1}{12}$. Note that some edges, notably AC and AB, will be painted more than once, because more than one ant travels along them.

Now we repeat the process. If we start all ants at point A, then each ant has two possible paths: the edge AB or the edge AC. Edge AB has pheromone level $\frac{1}{13} + \frac{1}{14} \approx 0.148$, while edge AC has pheromone level $\frac{1}{13} + \frac{1}{12} \approx 0.16$. Because edge AC has a slightly higher pheromone level, ants are slightly more likely to take this edge. And because *each* ant paints the trail it uses, the more ants that take the edge AC, the higher its pheromone levels will be, causing still more ants to follow it.

The net result is that, over time, edges on less costly paths tend to be more traveled. More importantly, since the level of the pheromone along a path is determined by the *total* length of the path, we are more likely to choose a suboptimal *step*, as long as it is on a more optimal *path*, something that mountain climbing would never do and simulated annealing would only do on occasion. Dorigo found that ACS produced better results than simulated annealing and genetic algorithms using far fewer trials.

9

〜〜〜〜〜〜〜〜〜〜

The Complete Saga

The fundamental problem of communication is reproducing at one point
either exactly or approximately a message selected at another point.

CLAUDE SHANNON
"A Mathematical Theory of Information"

In 1977, mission planners at NASA's Jet Propulsion Laboratory (JPL) in Pasadena, California, began planning a mission to Jupiter. The mission, ultimately christened *Galileo*, was originally scheduled for launch from the space shuttle in 1982. However, delays in the development of the space shuttle, and safety protocols put into place after the *Challenger* disaster (1986), meant that *Galileo* would not launch until October 18, 1989.

On April 11, 1991, mission controllers at JPL sent commands to the spacecraft to deploy its high gain antenna (HGA). However, the antenna stuck in half-opened position, and despite numerous attempts to fix the problem over the next two years, the HGA remained unusable. Engineers believe that the lubricant that allowed the parts of the antenna to slide past each other degraded during the years *Galileo* was in storage. The failure of the HGA meant that all the scientific data accumulated by *Galileo* would have to be sent via the low gain antenna (LGA). The problem is that the HGA would have been able to transmit data more than 3,000 times faster than the LGA.

Recall that the *bit* is the basic unit of information, corresponding to the answer to a single yes/no question. Conversely, using n bits allows us to distinguish between 2n different things. With 8 bits, we could distinguish between any of 28 = 256 different things.

The cameras on *Galileo* took pictures that were 800 pixels wide by 800 pixels high, for a total of 640,000 pixels. The (black and white) color of each pixel can be specified by a number between 0 (black) and 255 (white), with 254 shades of gray in between; specifying such a number requires 8 bits. Thus, a single black-and-white picture from *Galileo* would require 640,000 × 8 = 5,120,000 bits.

The HGA of *Galileo* had a transmission rate of 134,000 bits per second, so such a picture would have taken approximately 40 seconds to transmit; mission control had hoped to record about 50,000 pictures. Using the LGA, a single picture would take about 36 hours to transmit, and 50,000 would require more than 200 years. Unless something could be done, most of the data scientists hoped to acquire would be lost.

Galileo's problem wasn't unique to spacecraft but occurs anytime we transmit information from one place to another. For example, movies produce the illusion of motion by displaying a sequence of images in rapid succession. One of these images might consist of 2,000,000 pixels, each of which has a color specified by a 24-bit number; a standard frame rate would be 24 images per second. Thus, 1 second of a movie requires

$$24 \times 2{,}000{,}000 \times 24 = 1{,}152{,}000{,}000 \text{ bits}$$

As I write this, my cable modem tested at a download speed of about 25 million bits per second. This means that every second of a movie should have taken 46 seconds to stream.

Saving Spacecraft *Galileo*

By any measure, we should be drowning in digital media, streaming movies should be impossible, and the failed deployment of *Galileo*'s HGA should have been an unmitigated disaster. What saved *Galileo* and makes streaming videos possible is a *data compression algorithm*.

There are two types of compression. We can view these in terms of a message sender and a recipient. In *lossless compression*, the recipient can recover an exact copy of the sender's message. In *lossy compression*, the recipient cannot recover an exact copy but (provided the compression is of sufficient quality) can recover the essential details.

You might wonder why we would settle for lossy compression. As an example, consider the following: On December 17, 1903, Orville Wright made the first successful powered heavier-than-air flights. He sent a telegram to his father, Milton Wright, informing him of his success, concluding with "average speed through air 31 miles longest 57 seconds inform Press home Christmas."

Presumably, Orville did not talk or write this way in general. However, since the cost of a telegram is based on the number of words sent, Orville sent a compressed message rather than saying, "The average flight speed was 31 miles per hour, and the longest flight lasted 57 seconds; you should inform the press; and I'll be home for Christmas" (27 words), he conveyed this information in 13 words, a reduction of about 52%. We do something similar with text messages: "That's a very funny statement you just made and I'm laughing at it" collapses to "LOL." Here the reason isn't cost but speed: the former takes 66 characters to type, the latter 3, and at a compression of 95%, conveys the essential ideas.

Of course, the universe never gives us something for nothing. Losing information allows us to compress the data, but we must take pains to make sure the information lost is nonessential. Thus, Orville's message could have been shortened to "average 31 longest 57," but the new message would be ambiguous: was 31 the average length, duration, or speed? And "longest 57" might have meant the longest flight was 57 seconds, or 57 feet (or 57 yards). Moreover, when the compressed message is interpreted (*decompressed*), there is the possibility of introducing *artifacts*: things that were not in the original message but were generated because of the decompression. Thus, your comment might be met with "LOL," but this doesn't mean you'll do well as a stand-up comic.

To understand lossy compression, consider a party. The sounds of the party are produced by a number of generators: every partygoer, the music speakers, clattering dishware, and so on. To reproduce the sound of the party *losslessly*, we'd need to record, for every sound generator, the noise and volume of the sounds it produced at every point during the party. This requires a significant amount of data.

However, *most* of the sound comes from just a few generators: perhaps the music speakers and a few of the louder partygoers. If we keep only the data produced by these generators, and discard the rest, we'll have a *lossy* compression of the sound of the party. The art of lossy data compression centers around choosing the right volume for the right generators.

In 1986, computer scientists and telecommunications engineers realized that communicating large images electronically would be impossible without some form of data compression. The Joint Photographic Experts Group (JPEG) began meeting and produced a set of guidelines for image encoding. The result was the JPEG standard, based on the *discrete cosine transformation* (DCT). DCT required more computing power than was available onboard *Galileo*, so mission planners adopted a modified version of it known as the *integer cosine transformation* (ICT).

Both the DCT and the ICT work as follows. Consider a typical grayscale image, where each pixel is specified using an 8-bit number; this corresponds to assigning a number between 0 and 255 to each pixel. Suppose one row of an image corresponds to the sequence:

$$200 \quad 143 \quad 145 \quad 95 \quad 133 \quad 93 \quad 116 \quad 75$$

This row might appear as in Figure 9.1 (Original). We can view this as a vector with 8 components, each of which is specified by an 8-bit number.* This *bitstream* contains 64 bits; using *Galileo*'s LGA, it would take about 2 seconds to transmit.

Our goal is to try to reproduce this vector, or one very close to it, as a linear combination of a set of agreed-upon *basis vectors*. In terms of our party analogy, the basis vectors are the sound generators, and we have two tasks. First, pick a suitable set of sound generators. Second, determine how loud each sound generator should be set.

Since our image is an eight-component vector, we'd ordinarily need eight basis vectors to reproduce it exactly. Both DCT and ICT have a preferred set of basis vectors (as does a related approach, using *wavelets*), but to illustrate the key features of the method, we'll use the simpler set of basis vectors $[1, 1, 1, 1, 1, 1, 1, 1]$, $[1, 1, 1, 1, -1, -1, -1, -1]$, $[1, 1, -1, -1, 1, 1, -1, -1]$, and $[1, -1, 1, -1, 1, -1, 1, -1]$, together with four others, which, for

---

*Strictly speaking, a number like 95 can be expressed using 7 bits. However, the bitcount is based on the largest number that can be used. Since we *can* be required to handle numbers as large as 255, we *must* treat each number as an 8-bit number.

reasons discussed next, we can ignore. Remember that the dot product is a measure of how similar two vectors are to each other, and note that, in this case, the dot product of any two basis vectors is 0: they are as dissimilar as possible. We say that our basis is *orthogonal* (because if we view the vectors as directions in space, perpendicular vectors are as dissimilar as vectors can be). Our goal is to find a linear combination of the basis vectors similar to our original data.

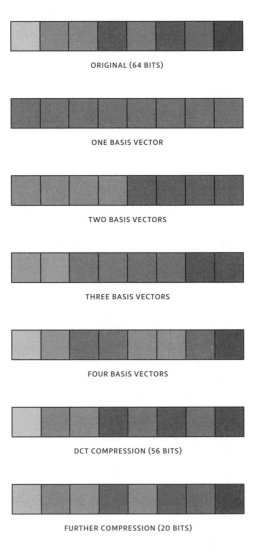

ORIGINAL (64 BITS)

ONE BASIS VECTOR

TWO BASIS VECTORS

THREE BASIS VECTORS

FOUR BASIS VECTORS

DCT COMPRESSION (56 BITS)

FURTHER COMPRESSION (20 BITS)

Figure 9.1. Original and Reconstituted Images

In other situations, we'd take one further step and *normalize* the basis vectors so the sum of the squares of their components is equal to 1. However, this requires working with nonintegers, which places an additional burden on computation; instead, we'll simply note that, for each basis vector, the sum of the squares of their components is 8.*

Now consider the dot product of our image vector [200, 143, 145, 95, 133, 93, 116, 75] with each of the basis vectors. The first four of these dot products will be 1,000, 166, 138, 188. These dot products give us the following information.

Remember our goal is to find a linear combination of the basis vectors that will be as similar as possible to the data. To begin with, our data consist of eight numbers. We might go further: all the numbers are positive, and the average value is 125. Thus, we might begin by approximating the data with a set of eight positive numbers whose average is 125. The simplest such set is [125, 125, 125, 125, 125, 125, 125, 125].

Suppose we only used our first dot product: we'd take 1,000 of [1, 1, 1, 1, 1, 1, 1, 1], then divide the result by 8. This gives us [125, 125, 125, 125, 125, 125, 125, 125]. In other words, if we use only the first dot product (1,000), we'd get a linear combination of one basis vector that approximates each data value with the average of all the data values. If we translate this vector into an image, we get Figure 9.1 (One Basis Vector).

Of course, this is a terrible reproduction. To improve it, we note that the actual data values *deviate* from the mean of 125; we can find the deviations by subtracting 125 from each, producing

75 18 20 −30 8 −32 −9 −50

Intuitively, if we can reproduce this sequence, then add it to the first, we'll obtain the data values. As before, we might consider these numbers. First, the sum of all eight numbers is 0 (this will always be the case for the deviations from the average value). Second, some of the numbers are positive and others are negative, and the negatives are concentrated in the second half of the sequence. Thus this sequence is similar to a sequence of positive and negative numbers whose sum is 0, and where the negatives are concentrated in the second half. In other words, it's like our second basis vector [1, 1, 1, 1, −1, −1, −1, −1].

*While it's not necessary to normalize our basis vectors, it's convenient if they have the same sum of squares.

If we multiply this basis vector by the next dot product (166), then divide by 8, we obtain [21, 21, 21, 21, -21, -21, -21, -21], which accounts for some of the deviations. It follows that if we use the first two dot products, 1,000 and 166, then the corresponding linear combination will be a better reproduction of our image: after dividing by 8 and rounding, we obtain [146, 146, 146, 146, 104, 104, 104, 104], and the corresponding image is shown in Figure 1: Two Basis Vectors. As you can see, this is a somewhat better reproduction of the original.

Once again, there is a difference between our original vector and the reproduced vector; we find the deviations to be

$$54 \quad -3 \quad -1 \quad -51 \quad 29 \quad -11 \quad 12 \quad -29$$

As before, some of these deviations are positive and others negative, but the average deviation is 0. If you break the sequence into two sets of four, then (very roughly speaking), the negatives will be concentrated in the second half of each set. Consequently, this set of deviations is similar to our third basis vector, [1, 1, -1, -1, 1, 1, -1, -1], and if we use our first three dot products, 1,000, 166, and 138, then the corresponding linear combination (after dividing the components by 8) will be [163, 163, 129, 129, 122, 122, 87, 87], and the corresponding image is shown in Figure 9.1 (Three Basis Vectors).

Notice that while using one or two basis vectors gave us a poor reproduction of the image, with three basis vectors we have something recognizably similar, and with four basis vectors (see Figure 9.1, Four Basis Vectors), the differences appear minor. This is where the compression possibilities come in: ordinarily, we'd use eight agreed-upon basis vectors to give an exact reproduction of the image; however, a reasonable reproduction can be found using just four. If we deem the reproduction sufficiently faithful, we can compress the eight values of the data into the first four coefficients of the linear combination: 1,000, 166, 138, and 188. These numbers form our compressed data and can be sent as 11-bit numbers.* This allows us to compress our bitstream from 64 bits down to 44 bits, a reduction of about 31%.

Where did the basis vectors come from? Since, in general, the re-

---

*Because the values can range from 0 to 2,040, where 2,040 is an 11-bit number.

maining deviations will include both positive and negative values, but the average deviation will be zero, we chose basis vectors whose components have similar properties. Moreover, it is convenient to have the basis vectors orthogonal, as this simplifies the calculations considerably. These factors led to our choice of basis vectors: you can verify that, except for the first, all have positive and negative values, and the components have an average of zero.

More generally, cosine is a mathematical function whose values over an interval are positive and negative with an average of zero; hence, the "C" in DCT and ICT. This provides us with a standardized set of basis vectors.* If we used these standard basis vectors on our data, we'd obtain the eight dot products

$$354 \quad 83 \quad 26 \quad 34 \quad 2 \quad 11 \quad -1 \quad 45$$

To account for both the magnitude and sign of the numbers, we'll need to record these as 11-bit numbers, so this data would require 88 bits to transmit, *more* than the 64 bits of the original data! This is an example of a *reverse compression*, where the "compressed" data take up more space than the original.

So how does DCT/ICT *compress* data? The answer comes from the "D" (or "I") part of the name. As a simple example, we might divide the numbers by 16 and round them off. This gives us

$$22 \quad 5 \quad 2 \quad 2 \quad 0 \quad 1 \quad 0 \quad 3$$

These numbers are 7-bit numbers, and so we've compressed our original data down to 56 bits, a 13% reduction. When decompressed, we'd recover

$$201 \quad 140 \quad 146 \quad 92 \quad 128 \quad 90 \quad 122 \quad 78$$

Again, we can compare the original to the reconstituted image in Figure 9.1 (DCT Compression), and see there is very little difference between the two. Moreover, if we'd wanted, we could gain even greater compression by using the first few dot products (for example, 25, 5, 2, and 2).

*There are actually several different sets of basis vectors. The decision of which set of basis vectors to use is not a mathematical problem, but a psychoperceptual one. See the epilogue.

Dictionary-Based Compression

Another approach toward data compression uses *dictionary coding*, a lossless algorithm. Suppose we want to compress the text ABRACA-DABRA. Computers store these characters using a sequence of 1s and 0s: thus, A is recorded as the sequence 01000001. Since there are eight digits, each of which is a 1 or 0, this is an *8-bit code*, and ABRACADABRA would require 11 × 8 = 88 bits.

How can we compress our data? One possibility is to assign a shorter code to A. For example, we might assign A the 3-bit code 001. This leads to a *dictionary code*, where each sequence corresponds to a specific character, symbol, or string.

There are two challenges we face when constructing a dictionary code. The first is that we don't know, in advance, how many different codes we'll need. To see why this is a problem, suppose we simply code each letter with its corresponding place in the alphabet: thus, A = 1, B = 2, and so on. ABRACADABRA would then be coded 1, 2, 18, 1, 3, 1, 4, 1, 2, 18, 1.

The problem is that, if we transmit this, we need to clearly delineate where one code ends and the next begins. The easiest way to do this, and the method adopted for Morse code, is to include a gap between successive symbols: 1-pause-2-pause-18-pause-1-pause, and so on. However, because network transmission rates vary, it's possible for a network slowdown to introduce a pause where none occurred. Thus, ABR might be *sent* as 1-pause-2-pause-18 but *received* as 1-pause-2-pause-1-pause-8-pause, which would lead to the incorrect decompression ABAH.

One solution is to require all codes to have the same length. Thus if we used two-digit codes, ABR would be 01-pause-02-pause-18. Even if we received 01-pause-02-pause-1-pause-8, the fact that *all* codes are two-digits would require us to read 1-pause-8 as 18, allowing us to correctly decompress the bitstream.

This leads to a new problem: there are only a finite number of codes of any given length. Thus we must decide, in advance, how many symbols we wish to encode. For example, suppose we wanted to code up to eight symbols. We can do this using eight 3-bit codes: 000, 001, 010, 011, 100, 101, 110, and 111. Processing the text ABRACADABRA, we generate the dictionary as follows.

The first symbol is A, so we assign it the first code: 000. The second symbol is B, which is new, so we assign it the next code: 001. The third symbol is R, which is again new, so we assign it the code 010. The fourth symbol is A, which we've already assigned code 000. The fifth symbol is C, which is new, so we assign it our fourth code 011. The sixth symbol is A, which already has a code. The seventh symbol is D, which is assigned the next available code 100. Proceeding in this fashion, we can code ABRACADABRA (nominally an 88-bit text) in 24 bits.

To improve our compression rate, note that while we have eight available codes, we only used five of them. Three of the codes, namely, 101, 110, and 111, weren't used. This suggests we might be able to use these codes to gain further compression.

A standard method of doing so is known as LZW, after its three developers: Abraham Lempel, Jacob Ziv, and Terry Welch. LZW emerged in 1984, after Welch introduced it as an improvement on earlier algorithms by Lempel and Ziv (known as LZ77 and LZ78, because they were published in 1977 and 1978). The GIF format, introduced in 1987 by CompuServe, uses LZW to losslessly compress images.*

As before, we'll assign codes as we go. But this time, rather than assigning codes to individual symbols, we'll assign codes to symbol *sequences*. In general, if we encounter a symbol or a sequence that we've already coded, we keep appending symbols until we have a symbol we *haven't* coded.

LZW begins with a base set of codes: in this case, we might use the codes 000 = A, 001 = B, 010 = R, 011 = C, and 100 = D. Now we process our text as follows. The first symbol is A. Since A has already been assigned a code, we join the next letter to produce the sequence AB. This doesn't have a code, so we assign this *sequence* to the next available code: AB = 101.

The next symbol is R, which already has a code, so we join the next symbol A. This gives us the sequence RA, which doesn't have a code, so we assign it the next available code: RA = 110.

The next symbol is C, which already has a code; joining the next symbol gives us CA, which receives the next code: CA = 111. This is also the last available code, so D will be coded as 100.

---

*GIF stands for "Graphics Interchange Format." Since "Graphics" has a hard G, purists insist that GIF also be pronounced with a hard G.

Now consider the next symbol, which is A. Rather than assigning it the code 000, we look at the next symbol, which is B, and recall that AB has the code 101. Likewise, the next symbol is R, which has a code, but begins RA, which also has a code 110. This allows us to code ABRACA-DABRA as

101	110	111	100	101	110
AB	RA	CA	D	AB	RA

for a reduction down to 18 bits.

Variable Length Codes

DCT/ICT and LZW all have one important advantage: they code as they receive the bitstream. Because they only require looking at the bitstream once, they are classified as *single pass compression schemes*.

Single pass compression techniques work quickly, but they might not compress a bitstream significantly. If we have the time, it's better to pass through the data once, to gather information about the distribution of the symbols, then pass through a second time to perform the actual compression.

For example, implementing a dictionary coding begins by deciding how long our code words will be; if our code words are n bits in length, then we'll be able to encode at most 2^n symbols. The risk is that, if we choose n to be too small, we might run out of code words before we run out of symbols. Thus, there are 26 letters of the alphabet, so we can use 5-bit codes (since $2^5 = 32$, which is more than enough). If we tried to use 4-bit codes, we could only encode $2^4 = 16$ different symbols. It's *possible* that a text might only use 16 of the 26 letters of the alphabet; however, if it uses 17 or more letters, the 4-bit codes will be insufficient.

In general, dictionary coding uses *longer* codes: thus, we might use 6-bit codes. It seems paradoxical to "compress" by substituting a 5-bit symbol with a 6-bit code; the reason that this works is that we will be able to take symbol sequences and replace them with 6-bit codes. For example, the letter sequence TH would ordinarily require 5 + 5 = 10 bits to send; if we can replace it with a 6-bit code, we will have reduced the message size by 4 bits.

We can improve our compression rate by passing through the data twice: first, to obtain information about the symbols actually used and then a second pass to code it. For example, suppose you know you're going to code a text. Since there are 26 distinct symbols in the alphabet, we need a 5-bit code.

But suppose you could look at the phrase first. The phrase might be

FADED CAB AGED BAD

where we'll ignore the spaces. There are only seven distinct symbols in this phrase: F, A, D, E, C, B, and G. Thus, we can get away with using a 3-bit code. We might use:

$$000 = F$$
$$001 = A$$
$$010 = D$$
$$011 = E$$
$$100 = C$$
$$101 = B$$
$$110 = G$$

Our message can then be coded in 45 bits.

We can go one step further by using a *prefix code*. Remember the reason we had to make all our codes the same length was that we couldn't guarantee the spacing between codes would be preserved in transmission. If we're using 3-bit codes, then there's only one way to read a transmission such as 000001010011010100001101001110011010101001010: we'd break it into the 3-bit codes and then consult our dictionary for the meaning of each code.

In a prefix code, no code begins a different code. This allows us to use codes of different lengths: for example, if we wanted to code four different things, we could either use four 2-bit symbols (00, 01, 10, and 11) or four prefix codes (such as 1, 01, 001, and 000). Because no code begins any other code, then even if the codes run together, it's possible to tell where one code ends and the next begins: 1001000011101 can only be read as 1, 001, 000, 001, 1, 1, 01.

An easy way to set up a prefix code is the following. Imagine a park with one entrance and several exits (Figure 9.2). Let the exits correspond to the different symbols we wish to code. Then draw a path from the one entrance that splits repeatedly until it reaches the exits. As long as the paths never cross, you can give directions to any exit by specifying a unique sequence of "left" and "right" commands.

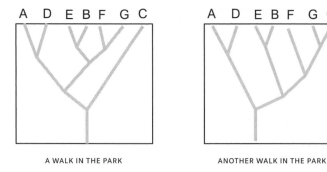

A WALK IN THE PARK   ANOTHER WALK IN THE PARK

Figure 9.2. Data Compression

For example, our park might look like the one on the left in Figure 9.2. To get to "A," we go left at the first fork, left at the next fork, and left at the third fork: we might describe this as "left, left, left." Likewise, the directions to get to "F" are left, right, right, left. If we represent "left" by 0 and "right" by 1, we obtain the following codes:

$$000 = A$$
$$001 = D$$
$$0100 = E$$
$$0101 = B$$
$$0110 = F$$
$$0111 = G$$
$$1 = C$$

Since the paths don't reconnect, no code will begin another code (otherwise, you'd be able to get to one exit *through* another exit). Thus, our bitstream becomes

0110000001010000110000101000011101000010101000001

and we've coded our original message in 49 bits.

We can decompress this as follows: imagine this as a set of directions through the park, where each "0" indicates "left" and each "1" indicates "right." The first digit of the compressed data is 0, so we go left at the first fork. The next digit is 1, so we go right at the next fork. The next digit is also 1, so we go right at the next fork. The next digit is 0, so we go left at the next fork. There are no further forks on this path, so it takes us to the exit: F.

Now we run around and go back to the park entrance. The next digit of the compressed data stream is 0, so we go left at the first fork. The next digit is 0, so we go left again. The next digit is 0, so we go left yet again, which takes us to the exit: A. In this way, we can recover the original message.

In 1951, Robert Fano (1917–2016), one of the pioneers of information theory, gave his students at MIT the choice of a final exam or a final project. One student, David A. Huffman (1925–1999), chose the project and was assigned the problem of data compression. It would not be an easy task: Fano himself was a pioneer in the field of data compression, and with Claude Shannon, had developed a then-standard method of data compression coding based on prefix codes. Huffman made little progress and was about to give up and take the final exam, when inspiration struck and he not only created a new method of data compression, but outdid Fano in one remarkable way: Huffman's approach is, under certain conditions, the best possible means of compressing a set of data.

Huffman's insight was the following: in the prefix code above, there's no relationship between the length of a code and the symbol it encodes. Thus, "F" has a 1-bit code, while "E" uses a 5-bit code. But since the code assignments are arbitrary, it makes more sense to assign shorter codes to more common symbols. This is in fact how Morse code works: "E" and "T," the most commonly used letter in English, are assigned "dot" and "dash," respectively, the two shortest signals. In "FADED CAB AGED BAD," we see that F, C, and G occur one time each; A and D both occur four times; E and B occur two times.

There are several ways to construct a Huffman code (and, consequently, the Huffman coding of a text can vary), though a simple one is this. First, we'll arrange the letters in order of their frequency and put a number of people at each exit, equal to the number of times the corresponding letter appears in the text. Thus, A has four people, D has four people, and so on (Figure 9.2, Another Walk in the Park).

Next, we'll create the park map as follows. At each stage, the two *smallest* groups of people walk together and join up. Thus, F, G, and C each have one person, so we might allow the persons at G and C to join up to form a group of two persons. Looking backward, this creates a forked path: to go *from* where the GC group has met *to* the G or C exit, we must take either the right path or the left path.

Now the F and the GC group have the fewest people, so they join up to form a FGC group. Again, this creates a forked path: one branch leads to the "F" exit, and the other to the previously created GC fork.

Since the FGC group has three people, but the B and E groups have two apiece, we join the B and E groups. Then, the BE-FGC group joins up, giving it a total of seven people. Now the two smallest groups are the A, D groups, so they join up to form the AD group with eight people; finally, the AD and BEFGC groups join, and everyone heads for the entrance. This produces the park map on the right in Figure 9.2, and the corresponding Huffman codes:

$$00 = A$$
$$01 = D$$
$$100 = E$$
$$101 = B$$
$$001 = F$$
$$1110 = G$$
$$1111 = C$$

If we use this assignment, our message "FADED CAB AGED BAD" can be compressed down to 39 bits as

001000110001111100101001110100011010001

Since the compressed data are a sequence of 1s and 0s, you might wonder if we can compress it again. The answer is . . . maybe. One way to evaluate how much compression we can perform is to find the informational entropy of the text.

Remember the informational entropy of a message corresponds to the number of yes/no questions required to recover the message. We can compute the informational entropy of our original text: FADED CAB AGED BAD has an entropy of about 2.57, which means that each symbol contains about 2.57 bits of information. As there are 15 symbols altogether, this message contains $15 \times 2.57 \approx 38.6$ bits of information. Thus, if we wish to code this message losslessly, we'll need at least 39 bits. Our Huffman code, which *did* use 39 bits, gives us the best possible compression.

10

\\\\\\\\\\\\\\\\\\\\\\\\\\\\\\

Complexity from Simplicity

To see the world in a grain of sand, and to see heaven in a wild flower,
hold infinity in the palm of your hands, and eternity in an hour.

WILLIAM BLAKE
Philosophical Essay on Probabilities

Mathematicians seek patterns in all things, even (and perhaps especially) that which seems capricious and unpredictable. During World War II, Lewis Fry Richardson (1881–1953) began a mathematical analysis of human conflict. One of Richardson's main results was that conflicts that produced k deaths occurred according to a *Poisson distribution*: murders ($k = 1$) occurred nearly every day, while decades would separate events like a world war ($k = 50,000,000$). This was a hopeful prediction in the postnuclear era, where fears of a global holocaust ($k = 1,000,000,000$ or more) abound: such events should be extremely rare.

Richardson's work may be seen as an extension of the problem confronted on a daily basis by the insurance industry: *why* something happens is of lesser importance than *how frequently* it will occur. In many cases, these frequencies are so predictable that the insurance company can set the premium for a policy without knowing anything else about the insured: flight insurance policies can be purchased at the airport for a nominal fee.

However, even better predictions (and more accurate premiums) can be found by identifying risk factors. Thus, a 25-year-old nonsmoker who exercises every day could be charged a lesser health insurance premium than an overweight 50-year-old chain smoker. In the same way, Richardson hoped to identify risk factors that led to international quarrels, in the hopes that this information could be used to prevent such quarrels in the future. Since most wars occur between neighboring countries, Richardson attempted to find whether there was a relationship between the length of the border between two countries and the likelihood of war between them.

Unfortunately, while it's easy to determine age, weight, and whether a person smokes, Richardson found it much harder to determine the length of national borders, because no two sources gave the same length! This became known as the *coastline paradox*.

Where borders are defined mathematically, like the western part of the US-Canada border (which, excepting some nineteenth-century surveying errors, runs along the 49th parallel), determining length is easy. But most borders are defined by natural frontiers: rivers, oceans, mountain ranges. So how can we measure them?

One strategy is the following. First, we need to choose the smallest length we can reliably measure. This will generally be determined by the physical size of our map: we can't *see* features smaller than a certain size, so we can't measure them. Next, mark points along our border separated by this shortest length and connect them with straight lines: this forms an approximation to the border itself. Finally, since it's easy to measure the length of a straight line, we can add up all the distances and find an approximation to the length of the border.

Why does this lead to the border paradox? Imagine a picture of an island like Great Britain. To measure the length of the coastline, we have to find points on the coast. But the smallest features our picture will be able to display will be 1 pixel across. Suppose each pixel represents a square, 1 mile across. Then the smallest length we can measure will be 1 mile. However, because the straight line is the shortest distance between two points, the corresponding points on the real coastline will generally be separated by a greater distance. Thus, if there are two points on the coast, 1 mile apart, they will be in adjacent pixels, and we'll measure the distance between them to be one mile. But since the actual coastline does not, in general, form a straight line, the actual distance between the two points will be greater than 1 mile. The measured length of the coast will in general be less than the actual length of the coast (Figure 10.1).

Figure 10.1. Real Coastline (Gray) vs. Actual Coastline (Dashed)

If we change the resolution, we'll change the size of the smallest features we can see. If our pixels correspond to a square 10 yards on a side, we'll be able to capture more of the features of the bay, and our measured length will increase. This will occur whatever distance each pixel represents, and so whatever our measurement for the length of the coast, it will *always* be less than the real length. Moreover, as our resolution increases, the measured length will also increase.

Might the measured length increase indefinitely, leading us to conclude that the coastline is infinitely long? This seems counterintuitive (hence, the paradox): after all, the country itself has a finite area, so how can it have an infinitely long coastline? Yet long before Richardson began looking at coastlines, mathematicians knew of curves with infinite length that enclosed finite areas. One of the simplest is known as the Koch Snowflake, based on work done by the Swedish mathematician Helge von Koch (1870–1924).

To produce the Koch Snowflake, we begin with a line of a specified length: for example, 1 meter (see Figure 10.2). We'll designate this Stage 1 Koch Snowflake. Next, we'll replace the middle third of the line segment with two line segments, each equal in length to the line replaced: this forms a "bump" in the line (Figure 10.2). The new figure is our Stage 2 Koch Snowflake.

We'll repeat this process: We'll remove the middle third of each line segment, and replace it with two line segments of equal length (Stage 3 Koch Snowflake). We repeat this process an infinite number of times to produce the actual Koch Snowflake (Stage 4 Koch Snowflake). However, just like our coastline, the fine details of the curve are too small to be displayed accurately in a pixellized image; this actually occurs around the eighth or ninth iteration.

Note that, at each stage, we take each straight line segment, break it into three parts, and replace it with four parts of the same length. Thus, the total length of the curve increases by a factor of $\frac{4}{3}$ every time. The Stage 1 Koch Snowflake has a length of 1 meter; the Stage 2 Koch Snowflake has a length of $\frac{4}{3} \approx 1.333$ meters; the Stage 3 Koch Snowflake has a length of $\frac{16}{9} \approx 1.778$ meters; the Stage 4 Koch Snowflake has a length of $\frac{64}{27} \approx 2.37$ meters. By the time we get to Stage 20, the Koch curve joins two points, 1 meter apart, with a curve whose length is approximately 237 meters! Since we repeat this process endlessly, Blake's infinity (in length, at least) in the palm of our hands becomes feasible.

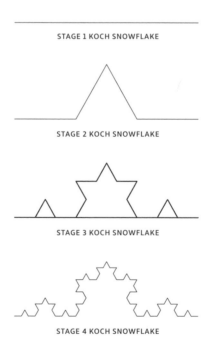

STAGE 1 KOCH SNOWFLAKE

STAGE 2 KOCH SNOWFLAKE

STAGE 3 KOCH SNOWFLAKE

STAGE 4 KOCH SNOWFLAKE

Figure 10.2. Space Filling Curve Koch Snowflake, Stages 1 through 4

Fractal Antennas

The Koch Snowflake is an example of a *space filling curve*. These curves have a number of useful properties, but one of the most important is that they can pack an enormous length of curve into a very small space. This is very important for any equipment that generates or receives radio signals.

Radio signals can be characterized either by their *frequency* or by their *wavelength*. The two are related as follows. Imagine standing next to a railroad track along which trains *must* travel at a speed of 30 meters per second (about 60 miles per hour). The length of each train car corresponds to the wavelength, while the number of cars passing by each second is the frequency. Because the speed is constant, knowing one allows us to compute the other. Thus, if we know each train car is 15 meters long, then we also know that two must pass by each second. Conversely, if we know that three train cars pass each second, it follows that each must be 10 meters in length.

Radio waves have roughly the same length as train cars, but travel 10 million times faster and have a correspondingly higher frequency. For example, a radio station might transmit radio waves at a frequency of 90.1 million waves per second: 90.1 megahertz (mHz), where the Hertz is named after the German physicist Heinrich Hertz (1857–1894). Because its transmission frequency is 90.1 mHz, the radio station is identified with FM channel 90.1.

Remember that information — whether it's music, news, or the traffic report — comes from reducing uncertainty. If the radio transmission *always* had a frequency of 90.1 mHz, then no information could be transmitted: you'd receive a single, invariable tone. It's necessary to alter some feature of the radio signal in order to transmit anything else. There are two ways to do this. The first is *amplitude modulation*: in our train analogy, it's as if all our cars were the same length but had different weights. If the train tracks ran over a scale, the varying weight of the train cars could then be converted into information: music, the traffic report, the latest news from the stock market. As the name suggests, AM radio uses this approach.

The other possibility is *frequency modulation*. In our train analogy, it's as if the *average* length of the cars was 15 meters, but the length of an individual car could vary. Again, as the name suggests, FM radio uses this approach.

Frequencies are important "property," in the sense that if someone is using a particular frequency for communications, then anyone else using the same frequency will cause interference. Thus, who can use which frequency is strictly regulated by government entities. In the United States, this regulatory authority is held by the Federal Communications Commission (FCC). For AM radio, the frequency is invariable, so AM frequencies can be spaced closely together: FCC regulations require a 9 kilohertz (kHz) separation, so if one AM station transmits at 1010 kHz, another must transmit at 1019 kHz or higher (or 1001 kHz or lower).

With FM radio, the frequency is an average, so the FCC requires a minimum separation of 0.2 mHz, with a 0.075 mHz variation in the frequency transmitted by a station. Thus, station 90.1, which transmits radio signals with frequencies between 90.025 mHz and 90.175 mHz, would not interfere with the broadcasts of station 90.3, which transmits between 90.225 mHz and 90.375 mHz.

Mobile telephone providers use even higher frequencies, in the range between 824 and 849 mHz. Even though these frequencies are very high, the speed of light is so great that the wavelengths are sizable: an 849 mHz radio signal has a wave length of about 30 centimeters, or a little more than a foot. This raises an important problem.

Imagine that you and a friend are standing on opposite sides of a passing train. Suppose you're throwing a hot potato, back and forth, at exactly the same speed as the train is moving, 30 meters per second.* Because the potato is hot, you can't hold on to it: you have to throw it as soon as you get it. If you throw the potato at a random time, it will probably hit the side of a boxcar. However, if you throw it at just the right time, it will pass through the gap between cars, and your friend will catch it. He'll throw it right back and once again, it's likely that it will hit the side of a boxcar.

In general, this experiment will result in nothing more than potato-smeared boxcars. But there are some combinations that will yield different outcomes. If each boxcar is 10 meters in length, your friend might stand 5 meters away. If you manage to throw the potato into the

*This is about the speed of a thrown football.

gap between two boxcars, and your friend throws it back to you, the potato will have traveled 5 + 5 = 10 meters. Since the potato is moving at the same speed as the train, the boxcars will have also moved 10 meters, which means that the potato will pass through the *next* gap between the boxcars.

Receiving a radio signal is a lot like catching the potato. As the radio signal passes through an antenna, it induces an electric field (the potato), which travels along the antenna and back at the speed of light. If the antenna isn't the right length, the field undergoes *destructive interference* by the radio signal: the potato hits the side of a boxcar. But if the antenna is the right length, there is *constructive interference* as the radio signal reinforces the electric field: the potato passes through the gap. As a result, the electric current running through the antenna gets stronger, and the radio signal is received.

As a general rule, an antenna should be half the wavelength of the radio wave it's intended to receive. For cell phones, this means antennas should be about 15 centimeters in length, which is about the size of the phone itself.

We can go further. If your friend stands half a boxcar length away, then a potato you throw between boxcars 1 and 2 will come back at you between boxcars 2 and 3. But if your friend stands a full boxcar length away, then the potato will still come back, this time between boxcars 3 and 4. In general, any multiple of half the boxcar length will allow potatoes to fly. Loosely speaking, the longer the antenna, the better the reception, and so cell phones *should* have antennas several meters in length.

Here is where the space filling curves are useful, because they allow us to pack a very long length into a very small space. On October 31, 2000, Nathan Cohen, physicist, radio engineer, and founder of Fractenna, received US Patent 6,140,975 for an antenna based around a space filling curve known as a fractal (a word whose origin we'll discuss later). Since then, a number of patents have been received by Cohen and others for different fractal antenna designs; Fractenna (now Fractal Antenna Systems) currently supplies a range of fractal antennas for public and private institutions.

You might wonder why we need to go through the trouble of producing a fractal, when we could simply coil the antenna into a spiral. The answer has to do with *self-similarity*. A fractal, like the Koch Snowflake, has the property that any portion of it resembles the original. Consider the Stage 4 Koch Snowflake (see Figure 10.2). Within it, you'll see portions that resemble the Stages 1, 2, and 3 Koch Snowflakes, just on a much smaller scale. A spiral doesn't have this property: zoom in on a spiral, and it will look like a section of a nonspiral curve.

How does this affect reception? Remember that the best length for an antenna is a multiple of half the wavelength of the radio signal it's intended to receive. However, different shapes have different reception properties. Imagine some shape to be particularly well suited for receiving a radio signal. The self-similarity of a fractal means that shape will be reproduced in many different scales, with many different lengths. As a result, the fractal antenna can not only pick up a range of frequencies but the reception over that range will be good.

Breaking the Dimensions

Space filling curves have other features worth noting. Since the smallest feature we can see in an image depends on the size of our pixels, at some point we're not able to display the features of the Koch curve. By Stage 20, the twists and turns of the Koch curve are about the size of individual atoms, and well before then details of the Koch curve are too small to be recorded in individual pixels.

This raises a new question. The Stage 20 Koch Snowflake will look like a version of the Stage 4 Koch Snowflake with somewhat thickened lines. But if we zoom in on what appears to be a thick line segment (▬), which appears to be a two-dimensional object with length and width, we'll see it resolve into earlier stages of the snowflake, and we would see the entire Koch Snowflake again, which consists of line segments (—), which are one-dimensional objects with length only. But even these have some width (they are at least one pixel across), and if we zoom in again, we see additional stages of the curve. We might ask: Is the Koch Snowflake a one-dimensional object or a two-dimensional object?

To answer this question, we need to determine what we mean by the dimensionality of an object. Consider the following: When we say that a point, a line, a plane, and a cube have zero, one, two, or three dimensions, what do we really mean? The question is so fundamental that it's very difficult to answer: it's like trying to explain the color "blue" to someone who does not have an innate conception of it. Intuitive concepts of dimension, such as the idea that a rectangle is two dimensional because it has length and width, break down when applied to objects like the Koch Snowflake.

Instead, we can try the following approach. Imagine an object like a line segment, and suppose we want to cover it using boxes (squares, in this case). We can use a single large square to cover it, but that is inefficient. Instead, we can use a multitude of smaller squares. If our covering squares are $\frac{1}{k}$ units on each side, and it requires N_k squares to cover the object, then the *box counting dimension* will be approximated by $\frac{\log N_k}{\log k}$, where the larger the value of k, the better the approximation.

For example, consider a square, 1 foot on each side. If $k = 4$, so our covering squares are $\frac{1}{4}$ foot on each side, it will take $N_4 = 16$ such squares. This suggests a box counting dimension of around $\frac{\log 16}{\log 4}$. If we double, starting from 1, we'd have to double 2 times to get 4, and 4 times to get 16, so $\frac{\log 16}{\log 4} = \frac{4}{2} = 2$.

Of course, $k = 4$ isn't very large, so we might try a bigger value. If $k = 16$, so our covering squares are $\frac{1}{16}$ foot on each side, we'll require $N_{16} = 256$ squares. This suggests a box counting dimension of $\frac{\log 256}{\log 16} = \frac{8}{4} = 2$ again. Or we could try $k = 64$, so our covering squares are $\frac{1}{64}$ foot on each side, and we'll need $N_{64} = 4{,}096$ of them, giving us a box counting dimension of $\frac{\log 4096}{\log 64} = \frac{12}{6} = 2$ once more. It appears that no matter how large k gets, the ratio $\frac{\log N_k}{\log k}$ will be 2, so we might conclude the box counting dimension of a square is 2. Formally, we say that the box counting dimension of the square converges to 2 as k *goes to infinity*.

This fits our commonsense intuition that the dimensionality of a square is 2. But while we *check* our mathematics by seeing whether it returns the expected answer, the value of mathematics emerges from its ability to be applied to many different situations. In this case, let's find the box counting dimension of the Koch Snowflake.

To begin with, note that at each stage, the segments of the snowflake are line segments $\frac{1}{3}$ as long as those of the previous stage. Thus, we might proceed as follows. At Stage 1, we'll take $k = 1$ and measure the Stage 1 snowflake using a square whose sides have length $\frac{1}{1} = 1$ meter; it will take $N_1 = 1$ such segments. At Stage 2, we'll take $k = 3$, so our square will have a side of length $\frac{1}{3}$ meters. Since each segment of the Stage 2 snowflake also has a length of $\frac{1}{3}$ meter, and there are four segments, we'll require $N_3 = 4$ squares to cover the snowflake.* This allows us to approximate the box counting dimension as $\frac{\log 4}{\log 3} \approx 1.262$. At Stage 3, we'll take $k = 9$, so our square has side length $\frac{1}{9}$ meters, and it will take $N_9 = 16$ such segments, which should give a better approximation to the box counting dimension: $\frac{\log 16}{\log 3} \approx 1.262$. And at Stage 4, we'll take $k = 27$, so our square has side length $\frac{1}{27}$ meter, and it will take $N_{27} = 64$ such segments, for our best approximation to the box counting dimension: $\frac{\log 64}{\log 27}$ ≈ 1.262. As all our approximations agree, we might conclude that, as k goes to infinity, the box counting dimension of the Koch Snowflake goes to 1.262. A similar approach gives us the dimension of the coast of the Great Britain — around 1.25.

Here we confront a rather bizarre situation: the Koch Snowflake and the coastline of Great Britain have a dimensionality somewhere between 1 and 2. In 1975, mathematician Benoit Mandelbrot (1924–2010) suggested the term *fractal* for such objects, to reflect their *fractional dimensionality*; redundantly, we speak of the *fractal dimension* of such objects.

Of course, coastal erosion and prevailing wind and tidal currents change the shape of a beach, so talking about *the* length of a coast or its fractal dimension is meaningless; in practice, we select a smallest feature size (say, 1 meter) and compute the length by picking points on the coastline separated by that distance.

On the other hand, consider more permanent features, such as roads. Suppose we want to compare two different roads: one along a flat plain and the other through mountains. The first road may be arrow straight, without any curves; the other twists and turns as it winds along the mountainside. Clearly, driving the first road is easier than driving the second, and accidents caused by loss of vehicle control will be less likely. Thus, the fractal dimension of a road can offer insight into the risk of driving along it. This is the basis of US Patent 8,799,035, assigned to the Hartford Fire Insurance Company on August 5, 2014, and developed by economists Mark Coleman, computer scientist Darrel Barbato, and mathematician Lihu Huang, who suggest that route

*Strictly speaking, we *could* cover the snowflake with just 3 squares of side ⅓ by taking in the so-called *global properties* of the curve. However, incorporating these global properties complicates the mathematics without changing the final result, so it's best to focus on the *local properties*, in which case we'd need 4 squares of side length ⅓ meter to cover the 4 segments of the Stage 2 snowflake.

complexity can be used to determine an auto insurance premium. This can be done before the trip, if the exact route is known, or on an ongoing basis, using GPS devices to record driving characteristics.

In this context, complexity has no specific definition but is a combination of factors that can be used to determine risk. At the simplest level, the length of a route is a measure of its complexity: an accident is more likely to happen on a thousand-mile journey than on trips across the street.* For this reason, insurance companies sometime offer discounts to low mileage drivers.

At a higher level, the fractal dimension of a road can be part of the measure of complexity. A related quantity, the *lacunarity*, measures the fraction of a region that is *not* filled by the fractal object. In both cases, the higher the fractal dimension or lacunarity, the more the road twists and turns, and the more care the driver must take to stay on it.

Both fractal dimension and lacunarity rely on covering the object with arbitrarily small boxes. However, because a road is a real object, sufficiently small boxes will eventually reveal that the road is a one-dimensional object and not an object with fractional dimension (lacunarity doesn't suffer from this problem). For example, the Stage 4 snowflake can be covered by 64 boxes, each $\frac{1}{27}$ meter on a side, which suggests a fractal dimension of $\frac{\log 64}{\log 27} \approx 1.262$. However, since the Stage 4 snowflake actually consists of straight line segments, each $\frac{1}{27}$ meters long, then decreasing the box size will not capture more detail. If each box was $\frac{1}{81}$ meter on a side, the Stage 4 snowflake could be covered by 192 boxes, suggesting a fractal dimension of $\frac{\log 192}{\log 81} \approx 1.1964$, and if each box was 1/27,000 meters on a side, we'd need 64,000 of them, suggesting a fractal dimension of around 1.085. As k goes to infinity, the value of $\frac{\log N_k}{\log k}$ for the Stage 4 snowflake converges to 1, the dimensionality of a nonfractal line segment.

A standard solution to such a problem in mathematics is to take the greatest possible value: in this case, even though the Stage 4 snowflake is a one-dimensional object, there is a box size that will yield a fractal dimension of 1.262, so when calculating the complexity of the route (and determining the premium to be paid by the insured), the road is

*The oft-cited statistic that most accidents are more likely to happen close to home ignores the fact that most *driving* occurs close to home. Thus, more surf-related injuries occur in Hawaii than in Nebraska, not because Hawaii is particularly dangerous for surfers, but because there are more surfers in Hawaii.

treated as an object with dimension 1.262. With GPS devices and vehicle monitoring systems, it would be possible to measure the complexity of a route as it's driven, and Coleman, Barbato, and Huang suggest that insurance companies might charge the insured a fixed premium and then offer rebates or credits to customers who use less complex routes.

Fractal Education

An unusual use of fractal dimension appears in US Patent 8,755,737, assigned to Pearson Education on June 17, 2014, and developed by William Galen and Rasil Warnakulasooriya. Computer-aided instruction can tailor a student's educational experience by presenting material at the exact pace at which a student learns. But how can this pace be determined? The answer is assessment.

Remember that education researchers divide assessment into two categories: formative and summative. The primary difference between the two is how the *incorrect* responses are used. For example, consider the problem "Evaluate: $\frac{3}{4} + \frac{2}{3}$." In summative assessment, the only thing that matters is whether a student got the correct answer: in this case, $\frac{17}{12}$. Answers like $\frac{5}{7}$ or $\frac{6}{12}$ or $\frac{5}{12}$ would be equally incorrect.

However, in a formative assessment, these questions would lead to different responses. Thus, the student who found the answer $\frac{5}{7}$ evidently obtained it by simply adding numerator to numerator and denominator to denominator. This would lead to one form of intervention, possibly a review of the basic concepts of what a fraction represents. In contrast, the student who answered $\frac{6}{12}$ might have done so by misreading the + as a ×, and this would lead to a different form of intervention (possibly after additional questions to determine whether this is in fact the problem). Finally, the student who answered $\frac{5}{12}$ appears to understand the necessity of having a common denominator but had some difficulties transforming each fraction into an equivalent one with that denominator, so a third type of intervention would be necessary. In a perfect world, every student in a classroom would receive the exact intervention necessary.

Imagine a student working his or her way through a topic such as the addition of fractions. In the initial formative assessments, the student's performance would be dismal. But as the student learned the topic more thoroughly, he or she would do better and better. At some point, the student is deemed ready to take the summative assessment, and a sufficiently high score on this last would be interpreted as student mastery of the material.

How can we judge whether a student is ready to take the summative assessment? The obvious solution is to use scores on formative assessments. For example, suppose the formative assessments are 10-question quizzes. We might judge a student ready to take the summative assessment when he or she has achieved an 80% overall quiz average.

Unfortunately, there are some problems with this approach. If the average includes the first few quizzes, which will tend to have lower scores, it will be very hard for the student to achieve the 80% overall average: a student who has five quizzes with a 0 would need to get a 10 on the next 20 quizzes to move their average to an 80%.

Instead, we might use a *moving average* of the last few quizzes. For example, our rubric might be that, if the average on the last five quizzes is an 80%, the student is ready for the formative assessment. However, even this might not tell the whole story. Consider two students, whose last 10 quiz scores are as follows:

Student A: 0, 1, 2, 5, 6, 8, 8, 7, 9, 10
Student B: 0, 5, 10, 2, 3, 10, 7, 9, 4, 10

If we used only the average of the five most recent quizzes, both students would be deemed ready to take the formative assessment. But Student A's performance shows a steady rise in scores, while Student B's record shows wild variations.

Galen and Warnakulasooriya suggest using fractal dimension as a way to measure these variations. Because each quiz is taken at a discrete point in time, and there is a definite sequence to the scores, the set of scores forms a *time sequence*. If we "connect the dots," the time sequence forms a progress curve, and we can find the fractal dimension of the curve, with higher fractal dimensions corresponding to greater variability and less predictability of the results of an assessment: thus even though the second student achieved an 80% average on the last five quizzes, his scores were so variable that his score on a summative assessment is essentially unpredictable. In contrast, the second student, who also averaged 80% on the last five quizzes, had much less variability, so if she were given a summative assessment, her score would more likely show mastery of the topic.

Cellular Automata

A fractal like the Koch curve is produced by repeatedly applying a simple rule to an object: Replace the center third of each line segment (–) in the curve with a wedge (^). This simple rule, repeatedly applied, produces an intricate pattern. A generalization of the idea of repeatedly applying a simple rule leads to the notion of a *cellular automata*.

Cellular automata emerged independently from the work of Stanislaw Ulam (1909–1984), who studied crystal growth, and John von Neumann (1903–1957), who studied the problem of self-replication. However, the most familiar example of a cellular automaton is one described by John Conway (b. 1937) and popularized by Martin Gardner (1914–2010) in *Scientific American* in October 1970.

Imagine a large grid like a chessboard. In a chessboard, we can distinguish between two types of squares: for example, black and white. Mathematically, we say each square has one of two *states*. In Conway's game of life, each square has one of two states: alive or dead. But unlike a chessboard, where the state of each square is fixed and unchanging, the squares in the game of life can change their state. It helps to think about discrete generations; we might see which squares are live or dead in one generation and then advance the clock and find the living or dead squares in the next generation.

There are just five rules:

1. A live square with zero or one neighbor dies (from "loneliness") by the next generation.
2. A live square with four or more neighbors dies (from "overcrowding") by the next generation.
3. A live square with two or three neighbors continues in the same state into the next generation.
4. A dead square surrounded by *exactly* three live cells becomes live in the next generation.
5. In all other cases, the state of a square remains unchanged.

For example, Figure 10.3 (Generation 1) shows an initial object (known as the *R-pentomino*). Squares B, C, D, E, and H have live cells in them, while the others do not.

What does the next generation look like? First, note that all of the unlettered squares are dead (empty) and have zero, one, or two

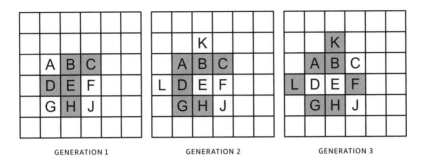

GENERATION 1   GENERATION 2   GENERATION 3

Figure 10.3. R-pentomino

neighbors, so by Rule 5, they remain dead. Next, square A is dead, but it's surrounded by exactly three neighbors (B, D, and E), so by Rule 4, it will become live in the next generation. A similar argument holds for square G. Meanwhile, squares B, C, and D are live and have two or three neighbors, so by Rule 3, they will continue to be live in the next generation. Finally, square E is live and has four neighbors, so by Rule 2, it will be dead in the next generation. This gives us the configuration of live and dead squares in Generation 2 (Figure 10.3, Generation 2).

We can apply the rules again. Thus, squares A, B, G, and H in Generation 2 are all live, with either two or three neighbors, so they remain live in Generation 3. D has four neighbors, so it dies by Generation 3; C has only one neighbor, so it dies as well. Meanwhile, squares K, L, and F are dead but have three live neighbors, so each has a live cell in Generation 3 (Figure 10.3). By a similar analysis, we can generate the configurations in Generations 4, 5, and so on, as far as we care to.

What if we change the rules? For example, under Conway's rules, a cell died if it had four or more neighbors. But what if it could survive more crowding, and only died when it had five or more neighbors? Or we can consider other variations: Conway counted diagonally adjacent squares as neighbors, while von Neumann did not. Still other variations are possible: thus we might take into account not only the number of neighbors but also the locations of those neighbors, so that a cell with three neighbors which are "above" it would continue to live, while a cell with three neighbors "below" it would die. The only possibility that we should exclude is *extinction*: all cells die, no matter how many neighbors they have.

Taking into account all of these possibilities, we find that Conway's rules are just one of many astounding

13,407,807,929,942,597,099,574,024,998,205,846,127,
479,365,820,592,393,377,723,561,443,721,764,030,073,
546,976,801,874,298,166,903,427,690,031,858,186,486,
050,853,753,882,811,946,569,946,433,649,006,084,095

possibilities! This number is far, far, far greater than the number of atoms in the universe, and if we just consider what these rules would generate from the R-pentomino, we could produce more patterns than we knew what to do with.

Most of these rules are uninteresting. For example, if no square ever died, and if all it took to turn a dead square into a live square was to have one live neighbor, then the board would quickly fill with live squares. Similarly, many rules lead to the extinction of all life, or to an ever expanding ring of living cells. But even if only 1 in a trillion rules allowed for continued survival, there would still be tremendous variability in the patterns produced. The problem is finding the 1 in a trillion that produces interesting patterns. The most amazing thing about Conway's game of life is that the simple set of rules Conway used *would* produce some very complex patterns.

The natural question to ask is, What *other* rules produce interesting patterns? The problem is that with so many possible rules, it's not feasible to examine more than the tiniest fraction of them. We need to reduce the complexity of the problem. One way to do so is to look at *one-dimensional cellular automata.*

These were studied by Stephen Wolfram (b. 1959). Imagine a line of squares, as long as we want, and consider any square and its neighbors to the right and left. The square itself might be alive or dead, and its two neighbors might be alive or dead, so there are eight possible states for the central squares. If we let 0 indicate a cell that is dead and 1 a cell that is alive, then these possible states for the central square range from 111 (the central cell and its two neighbors are alive) to 000 (all cells are dead). If a cell is in state 110, the central digit (1) indicates the cell itself is live; the digit to the left of the center (1) indicates that the square to the left of the center is also live; and the digit to the right of the center (0) indicates that the square to the right is dead.

Our rule can then be constructed by deciding which of these eight states lead to a live central cell (1) or a dead one (0). For example, we might have the rule

111	110	101	100	011	010	001	000
↓	↓	↓	↓	↓	↓	↓	↓
0	0	0	1	1	1	1	0

This indicates that if a cell and its two neighbors are alive (111), then in the next stage that cell will be dead (0); if a cell and the neighbor to its left are alive while the neighbor to its left is dead (110), then that cell will be dead (0), and so on.

If we always agree to list the starting states from 111 to 000, then the sequence of 1s and 0s completely describes our rule. Excluding the rule 00000000 (all cells die), we have 255 different possibilities for the one-dimensional game of life.*

Most of these rules lead to extinction or otherwise uninteresting patterns. However, as there are only 255 different possibilities, we can examine each one. For example, consider the rule 0001110. Suppose we began with an arbitrarily long line of cells, with a single live cell in the center:

$$0\ 0\ 0\ 0\ 0\ \mathbf{0}\ \mathbf{1}\ \mathbf{0}\ 0\ 0\ 0\ 0\ 0$$

Most cells would be dead, with dead neighbors: they'd correspond to the 000 state, and the central cell would remain dead.

However, there are three cells that would be different; we've put these in boldface to make them easier to see. First, there is the dead cell to the left of the live cell. This cell is dead, as is its neighbor to the left, but its neighbor to the right is the live cell, so it is in state 001. According to our rule, the central cell will become live (1) in the next step. Next, there is the live cell itself. Its two neighbors are dead, so it is in state 010; according to our rule, it stays live (1). Finally, there is the cell to the right of the live cell. This cell is dead, as is its neighbor to the

---

*We needn't stop here. If a cell's fate depended on the state of its two nearest neighbors, we'd have 32 possible starting states for each cell, from 00000 to 11111. Again excluding the extinction rule, this would give us 4,294,967,295 possible games of life.

right; its neighbor to the left is the live cell, so it is in state 100, so our rule makes this cell live in the next step.

While we could just show the new row of cells, showing which ones are alive and which are dead, it's customary to show all generations, so that we can see a history of our cellular automaton. To heighten the contrast between the live and dead cells, we'll omit the dead cells in the preceding generation (but keep them in the current generation to facilitate finding the next generation):

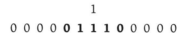

What happens in the next generation?

Most cells are dead with dead neighbors (000), so they remain dead. However, there are now five cells that are different; again, we put them in bold to make them easier to spot. The cell to the left of the leftmost live cell has state 001, and so it will be live (1) in the next round; the leftmost live cell has state 011, so it will also be live (1) in the next round. The central cell is in state 111, so it will be dead (0) in the next round.

The rightmost live cell is in state 110, so this cell will also be dead (note that our pattern isn't symmetric!). Finally the cell to the right of the rightmost live cell is in state 100, so it will be live in the next step. Showing our new generation, together with the previous two:

Applying our rule 0001110 for a few more generations gives us the history:

```
                  1
               1  1  1
            1  1        1
         1  1     1  1  1  1
      1  1        1           1
   1  1     1  1  1  1     1  1  1
   1  1     1              1     1
```

which we can continue as far as we'd like.

A Pleasant Walk, Interrupted

Cellular automata provide us with an easy way to generate complex patterns. One obvious use is for decorative patterns. However, there are other uses for complex patterns.

For example, consider the game of golf, where you hit a ball with a club in an attempt to land the ball in a hole some distance away. If the ball has no spin, the distance it travels (the *carry*) can be computed very precisely. However, it is generally impossible to hit a ball so that it has no spin at all. A spinning ball generates a force that alters its trajectory. The direction of the force depends on the direction of the spin relative to the ground. If the ball rotates backward (a *backspin*) relative to its direction of travel, the force pushes the ball upward, which increases the distance (*carry*) the ball travels. But with a frontspin, the force pushes the ball down and shortens its carry. When the axis of rotation is perpendicular to the ground (a *sidespin*), the ball will veer to one side or the other (producing a *fade, draw, hook,* or a *slice,* depending on the direction and magnitude of the curve).

All of this would be true if the golf ball were perfectly spherical. But this force can be changed if the surface of the golf ball is somewhat rough: thus, golf balls have *dimples*. A standard golf ball has between 300 and 500 dimples. The dimples have two main functions. First, the dimples disrupt the air flow around the ball, which helps reduce drag and allow the ball to travel farther. Second, if the ball is spinning, dimples increase the magnitude of the force generated.

As in life, more isn't necessarily better. If a player manages to impart a backspin, the increase in force allows the ball to travel farther, which is good. But if the player regularly imparts sidespins to the ball, the ball won't travel straight, which is undesirable. Fortunately, the laws that govern the flight of the ball are simple.*

In 1977, David Felker and Douglas Winfield, introduced the Polara golf ball, with a dimple pattern that helped reduced hooks and slices, regardless of the spin imparted to the ball. To do this, the dimple pattern was arranged so the ball has a definite "top" and "bottom," which

---

*Simple doesn't mean easy: 2 × 3 and 2348973 × 93418731984 are both simple, as they involve only multiplication.

neutralizes any lateral forces that drive the ball right or left. The Polara ball proved so successful at improving the games of amateur golfers that the United States Golfing Association (USGA) issued new rules that required golf balls meet certain symmetry requirements, which excluded the Polara ball.*

While the Polara ball, and similar golf balls designed to eliminate hooks and slices, are forbidden to tournament golfers, we might still look for dimple patterns that improve the flight characteristics of a golf ball. There's one problem. When designing the Polara ball, Felker and Winfield had a well-defined problem: prevent sidespin. By identifying a top, middle, and bottom segment of the golf ball, they could solve this problem using well-established rules of aerodynamics. But the USGA rules effectively prohibit such a golf ball. So the designer has no clear place to start.

One possibility is to use trial and error: draw a random dimple pattern on the ball, then simulate its flight characteristics using a computer. But what do we do if our random pattern gives the ball undesirable aerodynamic properties? We could choose another random pattern, but without some sort of guidance of what works and what doesn't, we have no hope that our second guess will be any better than our first. To be effective, trial and error needs some algorithmic way to alter our designs. One possibility is to use cellular automata. Hyoung-chol Kim of Kobe, Japan, developed such an approach, which became the basis for US Patent 8,301,418, assigned to SSI Sports on October 30, 2012.

Kim's work uses a two-dimensional cellular automata, like Conway's game of life, where the state of each location changes based on the state of its neighbors. However, in Conway's game and in Wolfram's one-dimensional versions, there are only two states for a location: live or dead. In Kim's version, each location has three states: inside, outside, or boundary.

Suppose we begin by covering the ball with a simple, basic dimple pattern: for example, 300 equally spaced dimples of identical size and shape. The dimples themselves form the locations with state "inside" (each dimple corresponding to one or more locations, in the way that a

---

*Nevertheless, many amateur golfers, who have no intentions of playing in tournaments, use the Polara ball.

US state covers several degrees of latitude and longitude); the undimpled portions correspond to the locations with state "outside," and the boundary between the two form the locations with state "boundary."

Next, we choose a set of rules for our cellular automata. These rules determine how the state of a location changes. For example, our rule might incorporate the rule "A location changes from 'outside' to 'boundary' if it's surrounded by 'boundary' locations on three sides."

Finally, we apply the rule and alter the dimple pattern. In this way, we can produce additional dimple patterns. What's important is that these dimple patterns are related to each other according to a simple rule. This allows us to apply a genetic algorithm as follows.

Suppose we begin with our simple, basic dimple pattern and determine its aerodynamic characteristics. By repeatedly applying the rules of the cellular automaton, we'll produce a set of related dimple patterns. It may turn out that our cellular automaton changes the dimple pattern in a way that worsens the golf ball's aerodynamic qualities. This suggests that the specific rule of our cellular automaton is *not* a rule we want to use, so we discard it.

However, it may be that the successive versions of the dimple pattern have better aerodynamic qualities. This suggests that our cellular automaton *is* a rule we want to use. By selecting the rules that improve the aerodynamic qualities, we can eventually produce an optimal rule to use for constructing golf ball dimples. Finally, we can begin with some random dimple patterns, apply our optimal cellular automaton rules, and produce a selection of designs to test.

11

∖∖∖∖∖∖∖∖∖∖∖∖∖∖∖∖∖∖∖∖∖

RSA . . .

If the theory of numbers could be employed for any practical and obviously
honourable purpose, if it could be turned directly to the furtherance of
human happiness or the relief of human suffering, as physiology and even
chemistry can, then surely neither Gauss nor any other mathematician
would have been so foolish as to decry or regret such applications.

G. H. HARDY
A Mathematician's Apology

In 1940, the English mathematician Godfrey Harold Hardy (1877–1947)
wrote *A Mathematician's Apology.* Hardy used the word *apology* in the
older sense of an explanation, not contrition. He wanted to explain why
he and others chose to study advanced mathematics. Among the reasons
Hardy found mathematics appealing was that it is, by and large, *useless.*

We might not view uselessness as a virtue, but Hardy had lived
through World War I and was writing at the start of World War II,
where the horrors of warfare had been magnified by technological and
scientific advances. But pure mathematics, because of its uselessness,
was blameless: "No one has yet discovered any warlike purpose to be
served by the theory of numbers or relativity, and it seems very unlikely that anyone will do so for many years."* While calculus could be

---

*Hardy, *A Mathematician's Apology,* 44.

used to build a bridge or a bomb, and statistics could help both epidemiologists and advertisers, pure mathematics — and number theory most of all — had no practical value: while it might not improve the world, neither would it cause harm.

Unfortunately for Hardy's equanimity, the illusion that relativity had no warlike purpose would be shattered by the atomic bomb. He might have taken solace in the fact that abstract algebra, another mathematical field he identified as useless, was instrumental in the breaking of the German military codes, though this would remain a closely guarded government secret until long after Hardy's death. But, had he lived, he might have been astonished to discover that number theory, which he identified as the most useless of all of areas of mathematics, would be the very underpinnings of twenty-first-century civilization.

Shopping from Your Bedroom

One of the fundamental features of life in the twenty-first century is that we can buy almost everything without leaving home: Each year, consumers spend more than a trillion dollars online. Moreover, the fraction of total sales made via the internet keeps increasing. In 2015, US e-commerce sales topped 7% of total retail sales. The trend is likely to continue, and many have forecast the death of the brick-and-mortar store within the next 20 years.

The ability to shop from home is not a new thing. Thus, in 1872, farmers throughout the Midwest received a one-page listing of items they could buy from Chicago-area merchant Aaron Montgomery Ward (1844–1913). While Ward was not the first to establish a mail-order business, he was far from the last. His primary competition came from fellow Chicagoan Richard Sears (1863–1914), who launched a mail-order business a few years later.

For more than a century, customers could select the items they wanted to purchase, submit an order to a retailer with their payment, and wait for their goods to arrive. The internet changed none of these features; the primary effect was to make the physical catalog obsolete. So why has the internet had such a profound impact on shopping?

What changed was how customers paid for their order. Consider the following scenario: Alice wants to buy something from Bob, so she sends Bob her credit card number. For convenience, we'll use a four-digit number: 3425. Bob receives the credit card number and charges Alice's account.

Now consider a third person who can listen in on the messages sent between Alice and Bob. Call this person Eve (for "eavesdropper"). Eve overhears Alice's credit card number and records it. Now Eve can buy things from Bob — or anyone else — and Alice gets the bill. Obviously, Alice doesn't want this to happen. So how can she prevent it?

First, Alice might communicate her credit card information to Bob by walking over to Bob, handing him her credit card, letting him take down the number, and then taking the credit card back. This method works — most of us do this on a regular basis — but what if Bob is in another state or another country? At some point *Bob* must communicate the credit card number to the credit card company. The customers of Ward and Sears relied on the security of the US mail system: to get credit card numbers, Eve would have to intercept mail passing between customers and retailers or between retailers and banks.

Rather than rely on physical security, Alice can *encrypt* her credit card information. In effect, she scrambles it so that only authorized persons can read it. As a simple example, Alice might reverse the digits of her credit card number before sending it to Bob. If her credit card number is 3425, she communicates 5243. Bob receives the number 5243 and reverses it to find 3425, Alice's credit card number. Eve, however, might overhear the number 5243, but since it's not Alice's credit card number, she can't use it.

Of course, once Eve knows that Alice encrypted her credit card number by reversing the digits, she can decrypt it the same way. If she intercepts 5243, she can recover Alice's credit card number by reversing the digits: 3425. We might try to keep the encryption method secret, but since Bob has to tell everyone how to encrypt their credit card numbers, Eve can find the method of encryption by posing as a legitimate customer.

Because the method of encryption (in this case, reversing the digits) gives Eve enough information to find the method of decryption (in this case, reversing the encrypted number), the preceding is an example of a *symmetric cryptosystem*. Another example of a symmetric cryptosystem is for Alice to increase each digit of her credit card by 1 (with 9s becoming 0s): if her credit card number is 3425, she would send the encrypted value 4536. Again, once Eve knows that this is the method of encryption, she can reverse it to recover Alice's card number.

Alice can make Eve's problem harder by keeping one vital piece of information secret. This piece of information is called the *key*. For example, suppose Eve knows that Alice adds *some* number to each digit, but doesn't know the actual number added. The number added is the key: in the preceding example, the key is 1, but if Alice used the key 2 (and added 2 to each digit), she would send (and Eve would intercept) the number 5647. Bob, who knows the key, could recover the credit card information; Eve, who does not, could not.*

This leads to the fundamental problem that must be solved before e-commerce becomes practical: every potential customer of a vendor must know the method of encryption. But in a symmetric cryptosystem, knowledge of the method of encryption gives you enough information to break the system. What is needed is *asymmetric encryption*, better known as *public key encryption*, where everyone knows how to encrypt a number, but that knowledge can't be used to decrypt the number.

Public key encryption exists because of a peculiarity of mathematics: It is often easier to *verify* a solution than to *find* a solution. For example, solving the equation $x^4 + 32x^3 - 8x^2 - 15x - 11.5625 = 440$ is difficult. But verifying that $x = 2.5$ is a solution is easy, since we can simply check to see whether $2.5^4 + 32(2.5)^3 - 8(2.5)^2 - 15(2.5) - 11.5625$ is equal to 440 (it is). This suggests a method of encrypting a number by producing an equation that it's the solution to.

Suppose Alice wanted to encrypt her credit card number 3425. We'll let 3425 be the solution to an equation. To avoid overcomplicating the problem, we'll make it the solution to an equation of the form $x^3 = N$.

How do we find N? Since we want $x = 3,425$ to be the solution to the equation $x^3 = N$, then it follows that $3,425^3 = N$. Thus Alice evaluates $3,425^3 = 40,177,390,625$ and sends this number to Bob. Bob can then recover Alice's credit card number by solving the equation $x^3 = 40,177,390,625$.

Of course, the encrypted value is useless unless there's a way to

---

*The alert reader might wonder how the *key* is communicated securely. We'll discuss this problem in chapter 12.

recover the original value. To solve this equation, Bob can use the Gold-ilocks approach. If one choice is too big, and another is too small, then something in-between will be just right. More formally, mathematicians speak of the *bisection method*.

It's helpful to break our equation into two parts: (1) the variable expression (x^3), which we can think of as a formula and (2) the constant value 40,177,390,625, which we can think of as a target. Bob's procedure is to evaluate the formula for different values in an attempt to hit the solution.

For example, Bob might make an initial guess of $x = 1,000$ for Alice's credit card number. He finds $1,000^3 = 1,000,000,000$, which falls short of the target value 40,177,390,624, so Alice's credit card number is not 1,000. Since $x = 1,000$ fell far short of the target, he might try a larger guess, say, $x = 5,000$. In this case, $5,000^3 = 125,000,000,000$, which is larger than the target value.

Because $x = 1,000$ gave him a number that was too small, and $x = 5,000$ gave him a number that was too large, Bob knows that the actual solution is somewhere between 1,000 and 5,000. Suppose Bob guesses $x = 3,000$. He finds $3,000^3 = 27,000,000,000$, which is still too small. But now Bob knows $x = 3,000$ is "too small," and $x = 5,000$ is "too large," so somewhere between the two is a number that is "just right." Again, if Bob guesses $x = 4,000$, he finds $4,000^3 = 64,000,000,000$, which is again too large, so the number that is just right is now between $x = 3,000$ and $x = 4,000$. With a little persistence, Bob can zero in on Alice's credit card number.

Unfortunately, Eve can do the same thing. The problem is that, while it's easy to write down an equation whose solution is the number we want to encrypt, the bisection method makes it easy for anyone to solve this equation. Therefore, we want to consider a situation in which we can't use this method.

Weekly Addition

The bisection method (and many others) rely on what's known as the *intermediate value theorem*. One of the consequences of the intermediate value theorem is that, if you add 1 to a number, you get a larger number. It follows that we might want to find a number system where adding 1 *doesn't* always increase a number.

Consider the expression 53 + 1. Arithmetic occurs when we give the answer: 54. Mathematics occurs when we identify *why* 54 is the answer. In this case, 54 is the answer because it's the number after 53. This suggests a way to generalize arithmetic: If $n + 1$ means the number after n, then we can extend the concept of addition to *any* system where "the thing after" makes sense. For example, consider the days of the week: "Wednesday + 1" would be "the day after Wednesday," and we could write Wednesday + 1 = Thursday. By a similar argument, we can write

$$\text{Monday} + 4 = \text{Friday}$$

since Friday is the fourth day after Monday. Now consider this expression Monday + 11. The eleventh day after Monday is Friday, and so we might write

$$\text{Monday} + 11 = \text{Friday}$$

Notice that adding 4 and adding 11 give us the same result. In some sense, 4 is like 11. Mathematicians write this as $4 \equiv 11$, which is read "4 *is congruent to* 11."

In weekday addition, the number 7 is of particular note, since the seventh day after any weekday is the same weekday:

$$\text{Monday} + 7 = \text{Monday}$$
$$\text{Tuesday} + 7 = \text{Tuesday}$$

and so on.

What does this say about 7? We might draw an analogy from ordinary arithmetic: Consider a sum like 2 + 0 or 8 + 0 or even 3,489,137,894 + 0. In all cases, adding 0 doesn't change the number: 2 + 0 = 2, 8 + 0 = 8, and 3,489,137,894 + 0 = 3,489,137,894. We say that 0 is the *additive identity*.

In the same way, we see that, in weekday addition, adding 7 doesn't change the day: adding 7 is like adding 0, so we can write $7 \equiv 0$. More importantly, 7 is the smallest (positive) number for which this is true. Thus, we say that 7 is the *modulus* (*mod* for short) of weekday addition.

The observation that 7 is like 0 has some important consequences. For example, since 14 = 7 + 7, and 7 is like 0, then 14 is like 0 + 0 = 0. Thus, $14 \equiv 0$. Similarly, since 18 = 14 + 4, and 14 is like 0, then 18 is like 0 + 4 = 4: thus, $18 \equiv 4$. Likewise, since $3 \times 6 = 18$, and 18 is like 4, then $3 \times 6 \equiv 4$. Perhaps the most unusual feature occurs with multiplication: we have $5 \times 7 \equiv 0$ (since 7 is like 0, then 5×7 is like $5 \times 0 = 0$).

There's one more useful feature. Our observation $18 \equiv 4$ means that, for the purposes of our system, 18 and 4 are the same thing. Thus, Monday + 18 will be the same weekday as Monday + 4. Since it's easier to work with a smaller number than a larger number, it's useful to *reduce the value by the modulus*. We say that 18 can be *reduced mod 7* to 4.

In mathematics, as in life, context is everything. All the preceding was based on our scenario of weekday addition. But if we considered clock time, we might start with $12 \equiv 0$, since adding 12 to any clock time gives the same clock time. This would give us $24 \equiv 0$ and $18 \equiv 6$ and similar statements.

You might object to this: Is $18 \equiv 6$ or $18 \equiv 4$? The answer is both — and neither; it depends on whether we're talking clock time or days of the week. To avoid confusion, we need to specify the modulus of the system. Thus, in the days of the week situation, the modulus is 7, and we say we are *working mod 7* and write $18 \equiv 4 \pmod 7$. In the clock time case, we are *working mod 12* and write $18 \equiv 6 \pmod{12}$. We could even choose our own modulus and work mod 83 or even mod 3,489,137,894 if we wanted.*

The significance of modulo arithmetic is that it's a system where the intermediate value theorem fails, which means that the bisection method (and many other methods of finding solutions to equations) won't work. For example, suppose Alice encrypted the digits of her credit card by cubing them, then reducing the number mod 1,000. Then the number 3,425 would be encrypted as follows:

1. As before, $3,425^3 = 40,177,390,625$.
2. We note that $40,177,390,625 = 40,177,390,000 + 625$.
3. Since we're working mod 1,000, then 1,000 is like 0, and so $40,177,390,000 = 40,177,390 \times 1,000$ can be reduced to $40,177,390 \times 0 = 0$.
4. Consequently, $40,177,390,625 = 0 + 625$.
5. Thus, $3,425^3 \equiv 625$.

*It's worth pointing out that $\pmod 7$ or $\pmod{12}$ applies to the congruence as a whole and not to one side. Think of it as a footnote: "$18 \equiv 4$, when we're working in a system where $7 \equiv 0$."

Now suppose Eve tries to use the intermediate value theorem to solve the congruence $x^3 \equiv 625$. She might find $5^3 = 125$ and $9^3 = 729$ and conclude that 5 is too small and 9 is too large, so she might conclude Alice's credit card number is between 5 and 9. Not only will she be looking in the wrong place, *no number between 5 and 9 will give her the correct value!* The intermediate value theorem fails in modulo arithmetic, and immediately removes one of the most powerful tools we have for solving equations.

Amazon's Primes

While modulo arithmetic makes it much harder for Eve to determine Alice's credit card number, the same is true for Bob. To make use of modulo arithmetic for encryption, we need to identify a crucial piece of information (the key) that will allow Bob to solve the problem *without* relying on the intermediate value theorem. One method goes back to seventeenth-century France, where Pierre de Fermat (1601–1665) made an interesting discovery about numbers.

To explain Fermat's result, we need to introduce some important ideas. Any number greater than 1 falls into one of two categories. First, we might be able to write the number as a product of two smaller numbers. Thus, $150 = 15 \times 10$. Any number that can be written as the product of smaller numbers is called *composite*. Second, it might be impossible to write the number as a product of smaller numbers. Thus, no pair of numbers smaller than 5 will multiply to 5. These numbers are called *prime*.*

Fermat observed that if p is a prime number and a is a number less than p, then $a^{p-1} \equiv 1 \pmod{p}$. For example, $p = 5$ is prime and $a = 2$ is less than 5. We see $2^{5-1} = 16$, and $16 \equiv 1 \pmod 5$. Likewise, $3^5 - 1 = 81 \equiv 1 \pmod 5$. To Fermat's frustration, this and similar discoveries were met with resounding silence.

About 70 years after Fermat's death, the Swiss-Russian mathematician Leonhard Euler (1703–1783) discovered Fermat's work and began extending it. One of Euler's results is the following. Suppose p, q are

*The classification of numbers into prime or composite specifically excludes the number 1, which is considered neither prime nor composite.

distinct primes, with $N = pq$. Then, for any a not divisible by p or by q, we have $a^{(p-1)(q-1)} \equiv 1 \pmod{N}$. For example, if we let $p = 3$, $q = 5$, then $N = 3 \times 5 = 15$. We can choose $a = 2$, and find $2^{(3-1)(5-1)} = 2^{2 \times 4} = 2^8 = 256 \equiv 1 \pmod{15}$. This discovery, now called the *Euler-Fermat theorem*, marked the birth of the branch of mathematics known as *number theory*. Results like the Euler-Fermat theorem are why Hardy and many others identified number theory as the epitome of mathematical uselessness: Of what use is this observation?

An answer would be given about 200 years later. In 1977, computer scientists Ronald Rivest, Adi Shamir, and Leonard Adleman, at the Massachusetts Institute of Technology, turned the Euler-Fermat result into the public key cryptographic system known as RSA, after their initials. In their original paper, published the next year, Rivest, Shamir, and Adleman described a communication between Alice and Bob, thereby starting a tradition that *any* discussion of cryptography, or communications in general, should include these two. To set up a RSA system, Bob goes through the following process:

1. He picks two primes p and q. For example, $p = 11$ and $q = 17$.
2. He computes $N = pq$. In this case, $N = 187$.
3. He computes $(p-1)(q-1)$. Here $(11-1)(17-1) = 160$.
4. He selects two odd numbers e and d whose product ed is 1 more than a multiple of 160. For example, we have $160 + 1 = 161 = 7 \times 23$, so Bob can use $e = 7$ and $d = 23$. There is a simple algorithm for finding these numbers, which we omit.
5. Bob then announces the *public modulus* $N = 187$ and the *public exponent* $e = 7$.
6. He keeps the *secret exponent* $d = 23$ somewhere safe.

Alice (or anyone else) who wants to send Bob a secret number m computes $m^e \pmod{N}$. Bob receives m^e and computes $(m^e)^d \pmod{N}$ and recovers the original message.

For example, suppose Alice wants to send the number 5 to Bob. Bob's public exponent is 7 and his public modulus is 187, so she finds $5^7 \equiv 146 \pmod{187}$ and sends the number 146 to Bob. Bob uses his secret exponent 23, and finds $146^{23} \equiv 5 \pmod{187}$, recovering Alice's number.

Why does this work? Remember that $187 = 11 \times 17$ and $(11-1)(17-1) = 160$. Thus, the Euler-Fermat theorem guarantees that, for any number

m not divisible by 11 or by 17, we have $m^{160} \equiv 1$, and so $m^{161} \equiv m$. In this particular case, what's important is that $5^{161} \equiv 5$. When Alice used Bob's public number 7, she found 5^7 and sent this value to Bob. When Bob then used his private exponent 23, he found $(5^7)^{23}$. But $(5^7)^{23} = 5^{161} \equiv 5$, allowing Bob to recover Alice's original number.

What if Eve tries to decrypt? Since she doesn't know Bob's secret exponent (in fact, the only person who should know this is Bob himself), she has to solve $m^7 \equiv 146 \pmod{187}$ for m. But without the intermediate value theorem, there's no good way to solve this congruence.

Could Eve determine the secret exponent? There is a straightforward algorithm for finding the secret exponent once we know p, q, the numbers whose product is N. But remember Bob doesn't tell us what p, q are; instead, he tells us what their product is. Thus, to find p, q, we have to find two numbers that multiply to 187. Equivalently, we must *factor* the modulus.

Most of us are familiar with factoring from school arithmetic, and from the problems we were given, it's natural to think that factoring is easy. But this is only because we've never been asked to factor anything truly difficult. To gain some insight into the magnitude of the problem, consider the problem of finding two numbers that multiply to 2,183.

Using a computer accelerates the process but only slightly. Roughly speaking, we can factor a number by forming a list of possible factors and then checking each of our possible factors to find one that works. The largest number we need to consider as a possible factor is less than or equal to the square root of the number. Thus, if we wanted to factor 2,183, our possible factors would be the numbers from 2 to $\sqrt{2183} \approx 46.7$. This gives us 45 possible factors: 2, 3, 4, and so on, up to 46.*

The problem is that the size of this list increases rapidly with the size of the number. Consider a number like 639,730,092,598,929,117,623. There are more than 25 billion potential factors. A computer that could check 1 million factors per second would take nearly 7 hours to go through them, before finding that this number is the product of 28,423,086,421 and 22,507,411,163 (which are both prime); Rivest,

---

*The actual number we'd need to check would be less, since every time a potential factor failed, it would remove several others. Thus, 2 will not be a factor of 2,183; this means it will be unnecessary to check 4, 6, 8, and so on.

Shamir, and Adleman noted that it would take a computer about a billion years to factor a 200-digit number (formed by multiplying two 100-digit prime numbers together). Most current RSA implementations use 300-digit numbers (formed by multiplying two 150-digit primes together), and 600-digit moduli are being considered.

Patentable Mathematics

Cryptographic systems such as RSA raise important questions about patentability. In particular, *could* RSA be patented? In order to receive a patent, an invention must satisfy several criteria, but the most important are that it be patentable, new, nonobvious, and useful.

Roughly speaking, only devices are patentable. Thus, mathematical formulas and algorithms are not, by themselves, patentable. In order to be patentable, the algorithm must be incorporated into some sort of device. While RSA itself could not receive a patent, a computer that implements RSA could be. Consequently, on December 14, 1977, they applied for a patent for a "Cryptographic Communications System and Method."

Even if the invention is deemed patentable, it must still meet the requirements of being new and nonobvious. However, the *newest* mathematics used in RSA predates the founding of the US patent office by nearly 30 years! And surely it is obvious to implement the algorithm on a computer.

What was new and nonobvious about RSA was the joining together of the different components. Thus,

- Choosing primes p, q to form $N = pq$ is not new.
- Finding e, d so $(a^e)^d \equiv a \bmod N$ is not new.

But joining these two to produce a cryptographic system was new. Consequently, Rivest, Shamir, and Adleman received US Patent 4,405,829 on September 20, 1983. Rivest, Shamir, and Adleman themselves went on to form RSA Data Security in 1982, which would be purchased by EMC in 2006 for $2.1 billion.

While the RSA patent has since expired (and Rivest, Shamir, and Adleman released it into the public domain before it expired), the Supreme Court has recently called into question whether such a patent is valid. In *Alice Corp. v. CLS Bank International*, decided June 19, 2014, the court ruled that running a known algorithm on a computer is not enough for patentability. Thus, the patentability of RSA would have rested on the question of whether connecting two known algorithms was sufficiently novel to warrant a patent.*

Improving RSA

RSA is now ubiquitous. An RSA implementation known as SSL/TLS (*secure sockets layer / transport layer security*) is used by any web page whose address is specified using "https" instead of "http." These web pages include those for online shopping and banking, as well as those required to log in to an email or social network account; they would not exist without some form of symmetric cryptosystem. It's not too much of a stretch to say that number theory created the twenty-first century.

However, while RSA is an extremely good cryptographic system, a number of practical considerations limit its usage to situations where keeping information secret is supremely important. The biggest problem is that encrypting numbers using RSA is computationally intensive; the SSL/TLS protocol used on "https" websites only uses RSA to exchange a key, which is then used to (symmetrically) encrypt the confidential data.

To understand the nature of this problem, we need to introduce the notion of the size of a number. In general, computer scientists measure the size of a number by the number of digits required to express it. Thus, 7 is a one-digit number, 23 is a two-digit number, 162 is a three-digit number, and so on.†

Note that 963 is also a three-digit number, so, in this sense, 162 and 963 are the "same size." As peculiar as this seems, there are a number of good reasons why this measure of size is useful. First, as far as a computation is concerned, the *value* of a number is unimportant: 2 × 3 is no more difficult than 9 × 7. In both cases, we perform *one* multiplication.

What about a multiplication like 162 × 963? If you do this computation the way you were taught in school, you'd perform 9 computations:

---

*In deference to Rivest, Shamir, and Adleman, the basic mathematics behind RSA is less well known among mathematicians than the basic mathematics behind Google's PageRank.

†Strictly speaking, the size is measured by the number of *binary digits* (*bits*) required to express it. One digit is equivalent to a little more than 3 bits.

1 times 9, 6, and 3; 6 times 9, 6, and 3; and 2 times 9, 6, and 3. There are some additional computations needed to find the final answer, but for convenience, we'll ignore these final steps.

Now consider Bob's situation. Alice sent him the encrypted number 146; to decrypt it, he must evaluate 146^{23}. This requires him to multiply 23 three-digit numbers together. First, Bob multiplies 146 × 146 = 21,316, which requires nine computations. Since he's working mod 187, he can reduce this to 185 before proceeding. Next, he needs to multiply 185 × 146 to obtain 27,010, taking another nine computations, but he can reduce this to 82 before proceeding. If Bob continues in this fashion, multiplying by 146 then reducing, he'll need to do about 200 computations to recover Alice's number.

What if Alice and Bob were using a real RSA system? In general, the product of the encryption and decryption exponents will be about the size of N, and the encrypted value will also be around the size of N. In practice, both Alice and Bob must work with numbers around the size of N. In a real-world RSA system, N would contain at least 300 digits. The product of the encryption and decryption exponents will also be a number with (around) 300 digits. We can form such a number by multiplying together two numbers, each with 150 digits, or by multiplying a number with 100 digits by 200-digit numbers, or any of a number of other combinations.

Suppose the public exponent were a 12-digit number, such as 989,195,963,183. To encrypt a number, Alice would need to perform around 300 *trillion* computations. A modern desktop computer can perform around a billion computations per second; it would take such a computer *more than three days* to perform this many calculations!

Of course, it doesn't take three days between the time you click Submit My Order and the confirmation. The reason is that over the years, computer scientists have found a number of algorithms that are far more efficient than the pencil-and-paper ones we're used to. One of the most important is known as the *fast powering algorithm*. Suppose we want to find 5^{25}. We compute

$$5^2, \left(5^2\right)^2 = 5^4, \left(5^4\right)^2 = 5^8, \left(5^8\right)^2 = 5^{16}$$

where we find each value by squaring the preceding value. Next, from the fact that 25 = 16 + 8 + 1, we can find $5^{23} = 5^{16}\, 5^8\, 5$. It may come as a surprise to discover that the fast powering algorithm predates computers by more than 2,000 years. Its first appearance is in the *Chandasutra* of Pingala (ca. 200 BC), an Indian mathematician.

The fast powering algorithm would reduce the problem of finding the product of $5^{989,195,963,183}$ significantly, since we could obtain all the necessary values by about 40 repeated squarings. Although each squaring will require multiplying two 300-digit numbers together (90,000 calculations), as opposed to multiplying a 300-digit number by a 1-digit number (300 calculations), the total number of calculations will be substantially reduced from 300 trillion to around 4 million.

The real problem occurs at Bob's end. Since the product of the public and private exponents will be roughly the size of N, a 300-digit number, it means that, if Alice is using a 12-digit public exponent, Bob must be using a 288-digit private exponent. Even with the fast powering algorithm, Bob must square Alice's number about 1,000 times, so he'll need to perform about 100 million calculations to recover Alice's number.

To be sure, modern computers can perform enormous numbers of calculations very quickly, and Bob will generally have more computational resources at his end than Alice has at hers. However, a secondary consideration is that each computation also requires some energy, which eventually winds up as heat: operating computers get very warm. Even if computational speed were not a problem, the danger of overheating a computer remains a driving force toward simplifying the algorithm.

Ancient Chinese Secrets

One approach to decreasing computational demands appears in US Patent 7,231,040, granted June 12, 2007, to Hewlett-Packard and developed by Thomas Collins, Dale Hopkins, Susan Langford, and Michael Sabin. Their patent is based on *multiprime RSA*.

The original RSA patent noted the possibility that the modulus N could be the product of three or more primes (and to "those skilled in the art," such an extension is obvious), so it covers RSA systems that use three or more primes. The Hewlett-Packard group was able to secure a patent on a particular implementation of a multiprime system.

Multiprime RSA (something of a misnomer, since RSA is already "multiprime") allows for gains in computational speed using the *Chinese remainder theorem algorithm*, so named because it solves a problem that first appeared in the *Mathematical Classics* of Sun Zi (fifth century?). Sun Zi's problem was:

> *There are an unknown number of things. If we count by threes, 2 remain; if we count by fives, 3 remain; if we count by sevens, 2 remain. How many things are there?*

In modern terms, Sun Zi is looking to solve a *system of simultaneous congruences*: to find the value of a number x for which all three of the following statements are true:

$$x \equiv 2 \ (\text{mod } 3) \quad x \equiv 3 \ (\text{mod } 5) \quad x \equiv 2 \ (\text{mod } 7)$$

Sun Zi's approach can be explained as follows. Suppose we start with a guess of $x = 0$. This fails all three congruences, so $x = 0$ is not the solution.

To proceed, we'll use trial and error. While we often deride trial and error as an inefficient way to solve a problem, it's often very effective, *provided* we have a way to modify our guesses to take us toward a solution. Indeed, the bisection method is a trial-and-error approach and works, as long as we have the intermediate value theorem to guide us.

What about the Chinese remainder problem? We know 0 is *not* a solution, so we need a way to modify it. The problem is that, if we change our guess, we'll change all three congruences. What we need is a way to isolate the effects of a change to just one congruence. The first important observation is that, if we increase our guess by a multiple of 3, it won't change the first congruence; if we increase our guess by a multiple of 5, it won't change the second; and if we increase our guess by a multiple of 7, it won't change the third. This means if we increase our guess by a multiple of 5 *and* 7, then the second and third congruences won't be altered. This is the key to the Chinese remainder theorem algorithm: If we're trying to solve a system of congruences, we can change one congruence at a time by adding the product of the remaining moduli.

Suppose we want to change the first congruence and solve $x \equiv 2$ (mod 3). We'll start with $x = 0$, which doesn't solve the congruence. The other two moduli are 5 and 7, so we add $5 \times 7 = 35$, and find

$$35 \equiv 2 \ (\text{mod } 3) \quad 35 \equiv 0 \ (\text{mod } 5) \quad 35 \equiv 0 \ (\text{mod } 7)$$

where we've solved the first congruence, namely, finding x so that $x \equiv 2$ (mod 3).

However, the other two congruences aren't solved. Since we're trying to solve the second congruence, we'll add the product of the other two moduli, 3 × 7 = 21, to our current solution to get a possible solution: 35 + 21 = 56. Since we've added 21, which is a multiple of 3, the congruence mod 3 will be unchanged. Since 21 is also a multiple of 7, the congruence mod 7 will also be unchanged. We find

$$56 \equiv 2 \ (\mathrm{mod}\ 3) \quad 56 \equiv 1 \ (\mathrm{mod}\ 5) \quad 56 \equiv 0 \ (\mathrm{mod}\ 7)$$

where, as promised, only the second congruence changed, though since we wanted a number congruent to 3 (mod 5), we're not quite there yet. So we add 21 again to get 56 + 21 = 77, which again leaves the congruences mod 3 and mod 7 unchanged and gives us

$$77 \equiv 2 \ (\mathrm{mod}\ 3) \quad 77 \equiv 2 \ (\mathrm{mod}\ 5) \quad 77 \equiv 0 \ (\mathrm{mod}\ 7)$$

which is still not what we want. But finally, if we add 21 again, we have 77 + 21 = 98 and find

$$98 \equiv 2 \ (\mathrm{mod}\ 3) \quad 98 \equiv 3 \ (\mathrm{mod}\ 5) \quad 98 \equiv 0 \ (\mathrm{mod}\ 7)$$

So 98 solves the first *and* second congruence.

Finally, adding 3 × 5 = 15 won't change the first or second congruence but will change the third. Again, 98 + 15 = 113, and

$$113 \equiv 2 \ (\mathrm{mod}\ 3) \quad 113 \equiv 3 \ (\mathrm{mod}\ 5) \quad 113 \equiv 1 \ (\mathrm{mod}\ 7)$$

which isn't what we want, but 113 + 15 = 128 gives us

$$128 \equiv 2 \ (\mathrm{mod}\ 3) \quad 128 \equiv 3 \ (\mathrm{mod}\ 5) \quad 128 \equiv 2 \ (\mathrm{mod}\ 7)$$

solving all three congruences simultaneously.

Finally, suppose we add or subtract a multiple of 3 × 5 × 7 = 105. Since this is a multiple of 3, the first congruence won't change; since this is a multiple of 5, the second congruence won't change; and since this is a multiple of 7, the last congruence won't change: thus, adding or subtracting 105 will change none of the congruences and will give us another solution. This tells us that 128 – 105 = 23 is the smallest positive solution.

Now consider this: 23 solves the system of congruences

$$x \equiv 2 \ (\mathrm{mod}\ 3) \quad x \equiv 3 \ (\mathrm{mod}\ 5) \quad x \equiv 2 \ (\mathrm{mod}\ 7)$$

But so does 23 + 105. If we think of $x = 23$ as a solution, then 23 + 105 is also a solution. This means that adding 105 is like adding 0 (in that it doesn't change anything), so we can view 23 as "living" in arithmetic mod 105. Consequently, we can say that $x \equiv 23 \pmod{105}$.

This is the *theorem* part of the Chinese remainder theorem: if the moduli are distinct primes, then we can always reduce a system of linear congruences to a single congruence whose modulus is the product of the original moduli. Thus, we began with a system of congruences with moduli 3, 5, and 7 and were able to rewrite it as a single congruence with a modulus of $3 \times 5 \times 7 = 105$.

Even more usefully, we can go backward. We can turn a single congruence into a system of congruences, where the product of the moduli is equal to the original modulus. Going backward is easy, provided we can factor the modulus. Thus, if we're trying to solve $x \equiv 23 \pmod{105}$, and we know $3 \times 5 \times 7 = 105$, then we find $23 \equiv 2 \pmod 3$, $23 \equiv 3 \pmod 5$, and $23 \equiv 2 \pmod 7$, so the corresponding system of linear congruences will be

$$x \equiv 2 \pmod 3 \quad x \equiv 3 \pmod 5 \quad x \equiv 2 \pmod 7$$

To use the Chinese remainder theorem to evaluate an expression like $146^{23} \pmod{187}$, we can think of the number 146^{23} as the "unknown number" in Sun Zi's problem: we want to solve $x \equiv 146^{23} \pmod{187}$. Remember Bob knows $187 = 11 \times 17$, so he can find x by solving

$$x \equiv 146^{23} \pmod{11} \quad x \equiv 146^{23} \pmod{17}$$

Of course, if Bob has to compute 146^{23} twice, we've made the problem worse! Fortunately, we're able to simplify both expressions considerably.

Consider the first congruence. Since we're working mod 11, we can make use of the fact that $146 \equiv 3 \pmod{11}$, so that instead of finding 146^{23}, we can find 3^{23}. In addition, we can make use of the Euler-Fermat theorem. Since 11 is prime, we know $3^{10} \equiv 1$, so $3^{23} = 3^{10} \, 3^{10} \, 3^3 \equiv 3^3 = 27$. But again, we're working mod 11, so $27 \equiv 5 \pmod{11}$. Thus, our first congruence becomes $x \equiv 5 \pmod{11}$.

By a similar argument, our second congruence can be reduced: first, $146 \equiv 10 \pmod{17}$ and then $10^{23} \equiv 107 \pmod{17}$, and we find $10^7 \equiv 5 \pmod{17}$. Thus, the problem of finding $146^{23} \pmod{187}$ becomes the problem of solving

$$x \equiv 5 \pmod{11} \quad x \equiv 5 \pmod{17}$$

The thoughtful reader may well find a simple solution to this congruence by inspection, but we'll tackle the problem like a computer and follow an algorithm. We'll start with an initial guess of 0, which doesn't work, so we'll add 17 (the product of the other moduli, in this case, there's only the one) repeatedly:

$$17 \equiv 6 \ (\text{mod } 11) \quad 17 \equiv 0 \ (\text{mod } 17)$$

$$34 \equiv 1 \ (\text{mod } 11) \quad 34 \equiv 0 \ (\text{mod } 17)$$

$$51 \equiv 7 \ (\text{mod } 11) \quad 51 \equiv 0 \ (\text{mod } 17)$$

$$\vdots$$

$$170 \equiv 5 \ (\text{mod } 11) \quad 170 \equiv 0 \ (\text{mod } 17)$$

which solves the first congruence. Then we'll add 11 repeatedly:

$$181 \equiv 5 \ (\text{mod } 11) \quad 181 \equiv 11 \ (\text{mod } 17)$$
$$192 \equiv 5 \ (\text{mod } 11) \quad 192 \equiv 5 \ (\text{mod } 17)$$

Finally, we can reduce this by $11 \times 17 = 187$ to find the smallest solution: $192 - 187 = 5$.

Using this approach, we can reduce the number of computations required for decryption dramatically: using the two 150-digit factors of our 300-digit number, we can reduce the number of computations from 100 million down to about 25 million.

The basic mathematics of RSA work even if N is a product of three or more primes (as long as the primes are distinct). If N is the product of three 100-digit primes, Bob only needs about 10 million computations, and if Bob used four 75-digit primes, he could decrypt Alice's numbers almost as fast as Alice can compute them.

There are other advantages besides speed. RSA relies on access to large prime numbers. But determining whether a large number is prime is difficult; consequently, there are "few" large primes. This leads to two serious problems.

First, the security of RSA relies on the difficulty of factoring the modulus. But setting up an RSA system begins by choosing two prime numbers p, q, then multiplying them to form the modulus N. If, by chance (and the fact that there are a limited number of large numbers known to be prime), you happen to pick two numbers that multiply to someone else's modulus, you have found a factorization, and *either* of you can break the other's system. A 2012 study, led by Nadia Heninger and Zakir Durumeric of the University of California at San Diego, suggested that roughly 1 in 130 RSA moduli were the same!

Two systems using the same key can be avoided through the simple expedient of checking to see whether anyone else is using the modulus you want to use, though this raises some ethical questions. If you discover the modulus you want to use is *already* in use by somebody else, then you know how to decrypt any messages sent to that other person. Should you then inform this person that his or her system is no longer secure?

Another problem occurs when two users pick the same prime number. Thus, suppose you pick the primes p, q and use modulus $N = pq$ and someone picks the primes q, r and uses modulus $M = qr$. The moduli are different, but they have a common prime divisor, namely, q. If someone can find the greatest common divisor of your two moduli, they can factor both numbers; Heninger and Durumeric found this occurred about 1 time in 200.

You might have been taught a method of finding the common divisor of two numbers by factoring. Of course, this requires factoring, which we've already identified as a hard problem. It may come as a surprise, but we can actually find the common divisor *without* factoring using the *Euclidean algorithm*.

Suppose you want to find the greatest common divisor of 140,201 and 40,349. Start by dividing the larger by the smaller, keeping track of the remainder:

$$140{,}201 \div 40{,}349 = 2, \text{ remainder } 2{,}041$$

Ignore the quotient; we don't need it. Now take the old divisor and divide it by the remainder, forming a new quotient (again, we can ignore it, since we don't need it) and a new remainder:

$$40{,}349 \div 2{,}041 = 9, \text{ remainder } 785$$

Again, we'll take the old divisor and divide it by the remainder:

$$2{,}041 \div 785 = 2, \text{ remainder } 471$$

Continuing this process:

$$785 \div 471 = 1, \text{ remainder } 314$$
$$471 \div 314 = 1, \text{ remainder } 157$$
$$314 \div 157 = 2, \text{ remainder } 0$$

The last nonzero remainder (157) will be the greatest common divisor. Since it is a *common* divisor, this tells us that both of our original numbers were divisible by 157, and we find:

$$140{,}201 \div 157 = 893$$
$$40{,}349 \div 157 = 257$$

which allows us to factor both numbers: $140{,}201 = 893 \times 157$, and $40{,}349 = 257 \times 157$.

Multiprime RSA addresses several problems. First, two users will have the same modulus when they choose the same primes, and it's harder for two people to choose the same three or four numbers than it is for them to choose the same two numbers. Second, there are more known small (150-digit) primes than there are known large (200-digit) primes, so users have a greater number of primes to choose from, making it even less likely they'll choose the same numbers. Third, even if they do choose the same primes, a third party still won't have enough information to factor the moduli: thus, if one person chooses three primes and produces modulus 629,821 and another chooses three primes and produces modulus 343,097, Eve might discover that both are divisible by 43, and so $629{,}821 = 14{,}647 \times 43$ and $343{,}097 = 7{,}979 \times 43$. However, she would still have to factor 14,647 and 7,979 in order to break the two systems.

"Three May Keep a Secret, If Two of Them Are Dead"

Cryptography is often described in terms of two communicating parties, with a third person trying to listen in on the conversation: Alice and Bob, who are trying to keep a secret from Eve. But in the real world, several people might try to share a secret. This complicates the problem of secrecy since, as Benjamin Franklin famously opined in 1735, "Three may keep a secret, if two of them are dead." In 1988, Johan Håstad of the Massachusetts Institute of Technology showed that, even if RSA itself was secure, multiparty RSA might not be.

Suppose we set up RSA with an encryption exponent of $e = 3$ and a modulus of 161. If we want to encrypt the number 2, then we find $2^3 = 8$, but $8 \equiv 8 \pmod{161}$. Thus, if we try to encrypt 2 as the number 8, Eve could simply solve the congruence $x^3 \equiv 8 \pmod{161}$ by solving the equation $x^3 = 8$. Since the intermediate value theorem applies to the equation, the equation can be solved easily, and the original value of the message can be recovered.

To avoid this, we need to guarantee that x^3 is greater than N. This can be done by *padding* the number. For example, $6^3 = 216$, which is greater than the modulus. Consequently, the cube of any number greater than 6 will be reduced, invoking the properties of modulo arithmetic and breaking the intermediate value theorem. This suggests we should add 6 to any number we wish to encrypt. Thus, to send the number "2," we'd first pad it to make it "8." Then we'd evaluate $8^3 = 512$. Since 512 is greater than 161, it will be reduced: $512 \equiv 29 \pmod{161}$. Since we've used the properties of mod 161 arithmetic, we've broken the ability to use the intermediate value theorem: recovering the message requires the attacker to solve the *congruence* $x^3 \equiv 29 \pmod{161}$.

Håstad's discovery was that, under the wrong conditions, multi-party RSA could be broken despite padding. Håstad's attack works as follows. Suppose Alice, Bob, and Charlie all set up RSA systems, using N_1, N_2, N_3 as their respective public moduli, and they all choose the same e for their public encryption exponents. As before, they keep their private decryption exponents secret. We'll assume N_1, N_2, and N_3 have no common factors (otherwise, the RSA systems are vulnerable to an attack based on the Euclidean algorithm). For illustrative purposes, suppose we have $N_1 = 161$, $N_2 = 209$, and $N_3 = 221$, and $e = 3$. Zachary wants to communicate the number "2"; he pads it, making it into 8 and then finds:

$$29 \equiv 8^3 \pmod{161} \quad 94 \equiv 8^3 \pmod{209} \quad 70 \equiv 8^3 \pmod{221}$$

He sends these numbers to Alice, Bob, and Charlie, respectively. Alice, Bob, and Charlie then decrypt the values by using their secret exponents.

Now suppose Eve intercepts all three numbers (29, 94, and 70). She might be able to do this because they're all being sent from the same source, namely, Zachary's computer. She knows that they are all from the same original number, so she looks to solve

$$c \equiv 29 \pmod{161} \quad c \equiv 94 \pmod{209} \quad c \equiv 70 \pmod{221}$$

In effect, she's looking for the value of x^3 *before* it was reduced modulo 161, 209, or 221. Eve uses the Chinese remainder theorem and finds that $c = 512$ is the smallest such value. Now she solves $x^3 = 512$, finding that

$x = 8$ was the message sent by Zachary. Finally, she removes the padding (which has to be publicly known, since otherwise legitimate users wouldn't be able to send encrypted messages) and recovers the secret digit "2."

Håstad's attack works primarily because the exponent is small. There are in fact a large class of *low exponent attacks* on RSA, leading to recommendations that the public exponent be at least 65,537. There's no real mathematical justification for the number 65,537; it's convenient, because it's easy to find the 65,537th power of a number through the fast powering algorithm: the sixteenth term in the sequence

$$a^2, \left(a^2\right)^2 = a^4, \left(a^4\right)^2 = a^8, \left(a^8\right)^2 = a^{16} \ldots$$

will be $a^{65,536}$. Assuming the public exponent is large enough and padding is done properly, multiparty RSA is secure.

Collaboration raises other problems. Suppose Alice, Bob, and Charlie all receive document D1 from Zachary and need to edit and revise it to produce a final document. Alice receives the document, decrypts it using her secret exponent, and makes some changes, producing document D2. *At the same time*, Bob and Charlie receive their copies, decrypt them, and make their own changes, with Bob producing D3 and Charlie producing D4.

Now Alice, Bob, and Charlie send their encrypted copies to everyone else. There are now three slightly different versions of the document floating around, which should be reconciled before proceeding. In the real (nondigital) world, this problem is easy to fix: If there's just one copy of the document, then only the person who has *physical* possession of the document can make changes. In the digital world, the problem is solved by storing the file at a central location, with protocols in place to make sure that the file can only be edited by one person at a time.

One way to set up the digital equivalent of a unique physical copy is to implement a *multiuser public key cryptosystem* based on file sharing. On July 20, 2010, Jesse D. Lipson received US Patent 7,760,872 for such a system. Lipson, a philosophy major, founded document sharing service ShareFile in 2005 and sold it to Citrix in 2011 for about $93 million in cash and stock.

As with single-user RSA, a multiuser RSA system would require each user have a public modulus N_i, though they'd announce the same encryption exponent e. This allows anyone to encrypt a file and upload it to a server. The encrypted file could then be decrypted by any of the authorized users.

However, there are some problems that must be overcome to set up such a system. If Alice wanted to set up a single-user RSA system, she'd pick primes and announce a public modulus N and a public exponent e. But not all values e are suitable as a public exponent: for example, if Alice chose $p = 7$, $q = 23$, giving her $N = 161$, then she'd need to find e, d where ed is one more than a multiple of $(7 - 1)(23 - 1) = 132$. Since 132 is a multiple of 3, $e = 3$ will *not* work as a public exponent (because one more than a multiple of 132 will *never* be a product of 3 and another number).

In a single-user system, this isn't a problem. Alice doesn't choose the public exponent 3. But in a multiuser system, Alice can't rely on the suitability of the public exponent. One possibility is to allow any of the authorized users to veto a proposed public exponent. However, this leads to another problem: *If we* know someone vetoed an exponent as unsuitable, we'd know something about that user's modulus. Thus, if Alice vetoed 3 as an unsuitable exponent, we'd know that 3 is a divisor of $(p - 1)(q - 1)$ for her values p, q.

Consequently, the public exponent must be chosen *first*. Then, the group members can choose p, q that will allow them to find a decryption exponent. This time, if they choose p, q that makes the public encryption exponent unsuitable, they would simply choose a different p, q.*

Suppose Alice, Bob, and Charlie pick $e = 7$ as the public exponent (this is far too small for security, but we'll use it for illustrative purposes). They select moduli 161, 209, and 221, and compute decryption exponents of 19, 103, and 55, respectively. As with single-user RSA, the public moduli are known. We also compute the group modulus, $161 \times 209 \times 221 = 7{,}436{,}429$. We'll need to use padding to avoid Håstad-style attacks; in this case, since our encryption exponent is 7, the smallest value we can encrypt will be 10 (since $9^7 < 7{,}436{,}429$). Thus, we'll add 10 to whatever we want to encrypt.

---

*Care should also be taken so that group members don't choose the same prime numbers. In this case, since group members have an incentive to maintain security, they can compare their moduli before announcing them to the general public.

Suppose Alice wants to encrypt the value 3. She adds $3 + 10 = 13$ and evaluates

$$55 \equiv 13^7 \ (\text{mod } 161) \quad 29 \equiv 13^7 \ (\text{mod } 209) \quad 208 \equiv 13^7 \ (\text{mod } 221)$$

Alice uses the Chinese remainder theorem to find a positive solution to

$$x \equiv 55 \ (\text{mod } 161) \quad x \equiv 29 \ (\text{mod } 209) \quad x \equiv 208 \ (\text{mod } 221)$$

which will be $x = 3{,}257{,}085$. She uploads this number to a website they can all access.

Now suppose Bob wants to edit the document. He takes the publicly accessible number 3,257,085, reduces it using $29 \equiv 3{,}257{,}085 \ (\text{mod } 209)$. He can use his own decryption key to find $13 \equiv 29^{103} \ (\text{mod } 209)$, then subtracts 10 to recover the encrypted value 3. Bob can make his edits (changing the number), re-encrypt, and upload. As long as download, decryption, encryption, and upload happen fast enough, Alice, Bob, and Charlie can collaborate in real time.[*]

What about Eve? Even if Eve knows the public moduli of Alice, Bob, and Charlie, she will be in the position of having to break a single-user RSA system. One consequence is that the common website doesn't have to be particularly secure. Even if Eve downloads 3,257,085 and uses Bob's public modulus to find $29 \equiv 3{,}257{,}085 \ (\text{mod } 209)$, she would need to solve $x^3 \equiv 29 \ (\text{mod } 209)$. Thus, this system is as secure as RSA itself.

What if Doug joins the group? Doug chooses public modulus 1,147 and finds his decryption exponent 463. Then one of the group members (it doesn't matter which one) retrieves the document and determines $735 \equiv 13^7 \ (\text{mod } 1{,}147)$. Solving the Chinese remainder theorem for the system that includes Doug's modulus,

$$x \equiv 55 \ (\text{mod } 161) \quad x \equiv 29 \ (\text{mod } 209) \quad x \equiv 208 \ (\text{mod } 221) \quad x \equiv 735 \ (\text{mod } 1{,}147)$$

gives $x = 62{,}748{,}517$ as the encrypted value to upload.

It's worth drawing an analogy to a physical system, where Alice, Bob, and Charlie all have keys (their private exponents) that open a door.

---

[*]In fact, all "real time" systems are based on this principle: a computer operates so much faster than the human beings who use it that it can switch its attention from one person to another and back again without the operators noticing.

When Doug joins the group, he gets his own key, and it's not necessary for Alice, Bob, or Charlie to get new keys.

However, a physical system has an important limitation. If we need to deny access to someone, we'd need to change the locks, which means changing *everyone's* keys. Lipson's approach allows us to deny access to one person, while at the same time leaving the other keys functional.

Suppose Charlie leaves the group. The uploaded value 62,748,517 must be altered so that Charlie can't read it. As before, we solve the system of congruences, but this time we'll omit Charlie's modulus:

$$x \equiv 55 \ (\text{mod } 161) \quad x \equiv 29 \ (\text{mod } 209) \quad x \equiv 735 \ (\text{mod } 1{,}147)$$

We find $x = 24{,}153{,}114$ as a solution, and upload this number. Now if Alice, Bob, or Doug download this number, they get the correct values for decryption, since

$$55 \equiv 24{,}153{,}114 \ (\text{mod } 161)$$
$$29 \equiv 24{,}153{,}114 \ (\text{mod } 209)$$
$$735 \equiv 24{,}153{,}114 \ (\text{mod } 1{,}147)$$

and so they can still use their private keys to decrypt the value. However, if Charlie downloads the number, he'd find

$$24 \equiv 24{,}153{,}114 \ (\text{mod } 221)$$

and if he used his private exponent, he'd find $80 \equiv 24^{103} \ (\text{mod } 221)$, which, after removing the padding, gives the (incorrect) decrypted value of 70. Thus, Charlie no longer has access to the encrypted data.* However, Alice, Bob, and Doug's keys still work, so unlike the physical system, the locks can be changed without requiring everyone to get new keys.

---

*In practice, as Lipson points out, Charlie may have stored older copies of the encrypted files. However, if the files are active and in the process of being changed, the new files will be inaccessible to Charlie.

12

〰〰〰〰〰〰〰〰〰〰〰

. . . Is Passé

"A slow sort of country!" said the Queen.
"Now, here, you see, it takes all the running you can do,
to keep in the same place. If you want to get somewhere
else, you must run at least twice as fast as that!"

LEWIS CARROLL (1832–1898)
Through the Looking Glass

During World War II, all sides communicated with their military commanders using sophisticated cryptographic systems, and all sides attempted to break the cryptographic systems of their enemies. When a group of British cryptanalysts, led by Alan Turing, successfully broke some of the German Navy's codes, the Allies were able to reroute convoys crossing the Atlantic to avoid German submarine patrols and helped Great Britain continue the fight against the Nazis.

Even more dramatic events occurred in the Pacific, where cryptanalysts of the US Navy had broken the Japanese codes. The architect of Pearl Harbor, Isoroku Yamamoto, sought to destroy the remaining ships of the US Pacific Fleet and set up an elaborate trap. By breaking the Japanese codes, the US Navy was able to set its own trap, and effectively destroyed the Japanese Navy at the Battle of Midway (June 3–7, 1942). The next year, on April 18, 1943, Yamamoto himself would be killed when the US Navy decrypted information about his itinerary during an inspection tour of the Solomon Islands.

These and similar incidents lead to grandiose claims that cryptographic successes shortened World War II by two or more years, an assertion that should be met with some skepticism: after all, *failure* of the cryptographic efforts at Bletchley Park and at Midway might have also shortened the war — by giving victory to Germany and Japan. Nonetheless, it should be clear that great efforts should be made to break cryptosystems — and equally great efforts should be undertaken to prevent them from being broken.

To date, existing attacks on RSA rely on features specific to an implementation. Thus, if the public exponent is too low, or several people choose a prime in common when constructing the public modulus, or the message to be encrypted is badly padded, a particular use of RSA can be broken, but security can be recovered by changing nonessential features.

However, there is always the possibility that a simple, efficient attack on RSA will be found tomorrow. Thus, cryptographers are looking for different methods of encryption, in the hopes that when and if RSA is broken, its replacement is waiting in the wings.

A Secret Recipe

The security of RSA is based on the difficulty of solving the problem $x^e \equiv c \pmod{N}$ for the unknown value x. The expression x^e, where x is an unknown value and e is a specific whole number is called a *polynomial*. Examples of polynomials include $x^5 + 12x^3 - 83x$, and in general anything that can be written with whole number powers of x, multiplied by specific constants, so a simple variant of RSA is to use a more general polynomial expression. However, this is so simple and obvious a change it would technically count as patent infringement (technically, because the RSA patent expired in 2000).

Instead, we might try an entirely different type of expression. Given a specific base a, we can form an *exponential expression* by raising the base a to some unknown power x. For example, 5^x, which indicates 5 is to be multiplied by itself some unknown number of times. Just as we can use the intermediate value theorem to solve the *polynomial equation* $x^3 = 512$, we can use the intermediate value theorem to solve the *exponential equation* $5^x = 3{,}125$. And just as working mod N eliminates our ability to use the intermediate value theorem, which makes

[248]

polynomial congruences harder to solve and allows us to create the RSA cryptographic system, working mod N makes exponential congruences harder to solve and allows us to create another cryptographic system.

Such a system actually predates RSA. On April 29, 1980 — more than three years *before* Rivest, Shamir, and Adleman received their patent for RSA — Stanford University was assigned US Patent 4,200,770 for what is now known as the Diffie-Hellman key exchange, invented by mathematician Whitfield Diffie, cryptographer Martin Hellman, and computer scientist Ralph C. Merkle.

Consider a manufactured food product such as mayonnaise. The law requires that the ingredients be listed on the label: soybean oil, egg yolks, distilled vinegar, water, sugar, salt, cider vinegar, spices. But the exact amounts are not listed. There's a good reason for this: If your competitors know the exact amounts of each ingredient, they can make your product and sell it for themselves. Since they don't, the best they can do is to make something that tastes *like* your product, and this difference is why my wife doesn't allow me to buy mayonnaise. This slight difference is the foundation for the trillion-dollar prepared foods industry.

However, even if a company can conceal the exact amounts of each ingredient from its competitors, it might seem impossible to conceal these amounts from those who produce the product. Thus, a competitor could break into a manufacturing facility and steal the recipe.

Surprisingly, we can secure our recipe from industrial espionage as follows. First, we set up two manufacturing facilities that produce the product. Next, send half of the recipe to one site and the remainder to the other. Each site prepares its half-recipe and then ships the product to the other, which adds the remaining half to make the product. In this way, *both* sites can manufacture the product, even though *neither* site has the complete recipe!

Of course, this is impractical when the product is a physical object and shipping costs are important. But when the object is information, shipping costs are negligible. Diffie-Hellman is based on such a "half recipe" scheme.

To implement the Diffie-Hellman key exchange:

1. Alice and Bob agree on a public modulus N (preferably prime, though this isn't an absolute requirement) and a public base a.

For example, $a = 5$, $N = 11$.

2. Alice picks a random number x, which will be her personal exponent, and evaluates $a^x \pmod{N}$. For example, she might choose $x = 3$ and find $5^3 = 125 \equiv 4 \pmod{11}$. This corresponds to the half-recipe (and so we'll call it Alice's *half key*), which she'll send to Bob.

3. Bob picks a random number y, which will be his personal exponent, and evaluates $ay \pmod{N}$. For example, he might pick $y = 2$ and find $5^2 = 25 \equiv 3 \pmod{11}$. This corresponds to Bob's half key, which he sends to Alice.

4. Alice takes Bob's number and raises it to her personal exponent x: $3^3 = 27 \equiv 5 \pmod{11}$. This combines Bob's half key with her own, and gives her the key.

5. Bob takes Alice's number and raises it to his personal exponent y: $4^2 = 16 \equiv 5 \pmod{11}$. Note that Bob now has the *same* key that Alice does, even though the key itself was never sent out!

Why does this work? When Alice raises Bob's number a^y to her personal exponent x, she is finding $(a^y)^x = a^{yx}$. Meanwhile, when Bob raises Alice's number a^x to his personal exponent y, he is finding $(a^x)^y = a^{xy}$. But since $yx = xy$, then $a^{yx} = a^{xy}$. Thus, like our secret recipe, Alice has combined Bob's half-recipe with her own half-recipe to obtain the key, while Bob can combine Alice's half-recipe with his half-recipe to obtain the *same* key. Moreover, Bob doesn't know what Alice's half-recipe contains, and Alice doesn't know what Bob's half-recipe contains. However, at the end of the Diffie-Hellman exchange, both Alice and Bob have the key in their possession.

Now consider Eve's problem. She knows that Alice sent Bob the number 4, and that Bob sent Alice the number 3. In order for her to find the key, she needs to know either x or y, which means she has to solve either $5^x \equiv 4 \pmod{11}$ or $5^y \equiv 3 \pmod{11}$. These are examples of the *discrete logarithm problem* (DLP), which is believed to be very hard to solve.

Zero Knowledge Proofs

Encryption solves one problem by ensuring that confidential information can't be read by unauthorized persons. However, there's another problem: ensuring that confidential information can't be misused by *authorized* persons. If Bob is legitimately in possession of Alice's credit card number, say, because she purchased something from him, what's to keep Bob from using Alice's credit card number for his own purchases? What is needed is some way to authenticate: to confirm that the person using the credit card is, in fact, the person who is authorized to use it.

In the real world, this authentication is done by signing the sales slip and (in theory, at least) comparing the signature on the slip to the signature on the card. But how can we perform such verification on-line? It doesn't do any good to provide additional information, like the CVV (Card Verification Value): all this shows is that you have physical possession of the card so, again, anyone who has your credit card can use it. Nor can we use the signature: a digital image of a signature can be replicated endlessly.

The Diffie-Hellman patent describes a way to use the system for authentication, and by extension, *any* public key cryptosystem can be used this way. As before, suppose we pick some number N and some base a. Suppose Alice wants to verify Bob's identity. First, Bob picks a secret exponent y and evaluates $a^y \bmod N$ and publishes this in some directory. Since the DLP is presumed hard to solve, no one but Bob knows the value of y; consequently, if Alice challenges Bob to prove his identity, he can reveal y, and Alice can confirm $a^y \bmod N$ is the value published by Bob in the directory.

Unfortunately, this will only work once, because as soon as Bob reveals y, anyone can use it and claim to be Bob. We might be able to delay the inevitable by encryption, but at the very least, *Alice* can impersonate Bob. What we need is some way for Bob to prove that he knows the value of y without revealing the actual value of y. Such a proof, which reveals nothing about the secret kept, is called a *zero knowledge proof*.

Remember the important feature of Diffie-Hellman is that it allows us to generate the *same* number in two different places without sending the number. Thus, we can use Diffie-Hellman as follows. Alice picks a secret exponent x and evaluates $a^x \bmod N$; she sends this to Bob. Bob, who knows the value of y, evaluates $(a^x)^y \bmod N$, and sends this value to Alice. Meanwhile Alice takes Bob's published number a^y, and since she knows x, she computes $(a^y)^x \bmod N$. If Bob is who he says he is, the number he sent her will be $(a^y)^x \bmod N$. In effect, Bob verifies his identity by generating the same key as Alice.

What's remarkable about this approach is that it allows us to do something that's impossible in the physical world: preventing those *legitimately* in possession of confidential information from misusing it. If Bob signs a credit card slip and gives it to Alice, then Alice, who has both Bob's credit card number and his signature, could use them to buy things for herself. But suppose Bob has authenticated himself to Alice using a zero knowledge proof. What happens if Alice tries to impersonate Bob in a conversation with Charlie?

To verify "Bob's" identity, Charlie picks a random exponent z, evaluates $a^z \bmod N$, and sends it to Alice; meanwhile, Charlie uses Bob's published value a^y and computes $(a^y)^z \bmod N$. Alice, who only knows her own personal exponent x, can't use it, since it would be a fantastic and unlikely stroke of luck to have $(a^y)^x$ equal to $(a^y)^z$. Even though Bob has verified his identity to Alice, Alice can't use this information to impersonate Bob to anyone else. As promised, Bob has proved his identity to Alice, but Alice has gained zero knowledge from the proof.

The limitation of Diffie-Hellman is that it can't be used to send a specific number. It isn't a cryptographic system. However, it can be used to exchange the key for a cryptographic system. In 1985, Taher Elgamal, then at Hewlett-Packard, invented a cryptosystem generally known as ElGamal (sometimes El Gamal or El-Gamal) based on Diffie-Hellman.

ElGamal is based on the existence of the *multiplicative inverse* (mod N): given some number p, find a number q so that $pq \equiv 1 \bmod N$. For example, if $p = 5$, then the multiplicative inverse (mod 11) will be 9, since $5 \times 9 = 45 \equiv 1 \bmod 11$. In general, the multiplicative inverse of a given number can be found easily and quickly using the Euclidean algorithm. The decryption exponent d for an RSA system is the multiplicative inverse of the encryption exponent e, mod $(p - 1)(q - 1)$.

Suppose Alice wants to communicate the number 8 using ElGamal. She and Bob go through the following steps:

1. Alice and Bob implement the Diffie-Hellman key exchange, and settle on the key 5.
2. To send the number 8, Alice computes $8 \times 5 = 40 \equiv 7 \pmod{11}$, and sends the result (7) to Bob.
3. To decrypt, Bob finds the multiplicative inverse of the key (9).
4. Bob multiplies Alice's number (7) by the multiplicative inverse of the key (9) to get Alice's number: $7 \times 9 = 63 \equiv 8 \pmod{11}$.

Elgamal (the person) did not apply for a patent, because Stanford

University argued that, while no cryptosystem was described in the Diffie-Hellman patent, ElGamal (the cryptosystem) was an obvious extension to "those skilled in the art." However, it's unclear what would have happened if it came to a legal battle.

More importantly, ElGamal became part of PGP (Pretty Good Privacy), invented by Phil Zimmerman in 1991. A few years later, PGP became the focus of a criminal investigation. The US government classified the cryptographic system as a munition and prohibited its export. Zimmerman argued that because PGP was a software package, its distribution was protected by the First Amendment right to freedom of expression. He was successful, and the government eventually abandoned the criminal investigation without filing charges.

There are several useful features of ElGamal. First, RSA encryption and decryption rest on raising the value m to a very high power, and this must be done for every message. This requires a great many computations, slowing implementation of RSA. In contrast, ElGamal relies on raising the public base to a very high power *twice* and then encrypts by multiplying the key by the message value; decryption relies on multiplying the resulting value by the multiplicative inverse of the key. The computational requirements are much lower, and, consequently, an ElGamal encryptions system runs much faster than a RSA system. From a practical perspective, this means that ElGamal can be used on devices with much less computational power. This is becoming increasingly important as we move toward an *internet of things*: if home appliances can be programmed via your cell phone and can interact with other devices, the day is rapidly approaching when coffeemakers and refrigerators will need sophisticated encryption systems; otherwise, someone could hack your coffeemaker and have it brew decaf, or have your refrigerator order a half side of beef.

Second, suppose Eve somehow obtains the key for the cryptographic system in use. For example, Bob sets up an RSA system with public modulus $N = 187$ and public exponent $e = 7$; he keeps private exponent $d = 23$ secret.

As a general rule, nothing can be kept secret forever, so Bob should assume that, sooner or later, Eve will discover the private exponent. Bob needs to periodically change the parameters of his cryptosystem. As we saw in the last chapter, every time we choose a modulus, we take a risk that someone else has already chosen that modulus, so it's best if we change the public exponent instead. Thus, Bob changes his public exponent to $e' = 17$, which makes his new decryption exponent $d' = 113$.

Now suppose Eve somehow discovers $d' = 113$. She can now read every message Bob sent using the encryption exponent $e' = 17$. Moreover, *she can read all of Bob's old messages as well.* The old messages used the encryption exponent $e = 7$; Eve can find a decryption exponent through the following steps:

1. She computes $e'\,d' - 1$, using the public/private exponent pairs she knows: $17 \times 113 - 1 = 1{,}920$.
2. She finds the multiplicative inverse of the original public exponent mod 1,920: this works out to be 823, since $7 \times 823 \equiv 1 \;(\text{mod}\,1{,}920)$.
3. She can use $d = 823$ as the decryption exponent for $e = 7$.

You might notice that $d = 823$ is not Bob's original decryption exponent, so you might wonder why this works. Remember for $N = 187$, an encryption/decryption pair e, d must satisfy $ed \equiv 1 \;(\text{mod}\,160)$. We can verify that $e = 7$, $d = 23$ works, but $e = 7$, $d = 823$ also satisfies $ed \equiv 1 \;(\text{mod}\,160)$, so they will work as an encryption/decryption pair!

The situation is even worse, because Eve can do this anytime Bob changes his public exponent. Once Eve knows a single encryption/decryption pair, she can find the decryption exponent for any encryption exponent Bob creates. Thus, a single compromised RSA key can reveal Bob's entire communication history — past, present, and future!

In contrast, consider our ElGamal system. Suppose that, in the first session, Alice and Bob choose 15 and 23, respectively, as their personal exponents. Alice sends Bob $5^{15} \equiv 13 \;(\text{mod}\,83)$, while Bob sends Alice $5^{23} \equiv 19 \;(\text{mod}\,83)$. Alice raises Bob's number 19 to her personal exponent 15 and finds $19^{15} \equiv 76 \;(\text{mod}\,83)$. Meanwhile, Bob raises Alice's number 13 to his personal exponent 23 and finds $13^{23} \equiv 76 \;(\text{mod}\,83)$. Now they're both in possession of the encryption key for the session: 76. This key should only be used for this session: it's called an *ephemeral key.*

In a later session, Alice might send Bob $5^{11} \equiv 55 \;(\text{mod}\,83)$, and Bob might send Alice $5^{13} \equiv 47 \;(\text{mod}\,83)$; Alice would then find the encryption key $47^{11} \equiv 45 \;(\text{mod}\,83)$, and similarly Bob would find $55^{13} \equiv 45 \;(\text{mod}\,83)$ as well.

Now consider Eve's problem. If she discovers Alice and Bob are using the key 45, this won't help her decrypt messages from the first session, which used a different key. Alternatively, if she solved the DLP for the first session and found Alice's private exponent, this wouldn't help her decrypt the messages in the second session. Unlike RSA, a single compromised ElGamal key reveals the contents of just one message.

Elliptic Curves

The difficulty of solving the polynomial congruence $x^e \equiv c \pmod{N}$, which would break RSA, and the exponential congruence $a^x \equiv k \pmod{N}$, which would break Diffie-Hellman and ElGamal, can be greatly reduced if there is an efficient way to factor large numbers, and currently it's believed that it's very hard to do so. But a quick, efficient method may be just around the corner. Thus, cryptographers consider systems whose security doesn't rely on the difficulty of factoring. One such system uses an *elliptic curve*.

Consider the following problem: How do we specify the location of an object? A standard approach begins with an agreed-upon origin and then specifies the location of a point by giving a set of directions for getting there. For example, "Start at the Empire State Building, go 3 blocks east and 8 blocks north" (which takes you to the Chrysler building); or "Start at the Empire State Building, go 3 blocks west and 7 blocks north" (which takes you to Penn Station) and so on.

Mathematicians refer to the individual directions ("3 blocks west") as *coordinates*, and the combined set ("3 blocks east, then 8 blocks north") as a *set of coordinates*. As long as we give the directions, we can list the coordinates in any order: "3 blocks east, 8 blocks north" describes the same location as "8 blocks north, 3 blocks east." However, if we omit the directions, we have to agree on an order, and, conventionally, we give the horizontal coordinate, followed by the vertical coordinate, with the whole enclosed in parentheses: thus "3 blocks east, 8 blocks north" would have coordinates (3, 8), where we take the east-west line as horizontal and the north-south line as vertical.

We can define a set of points by requiring the coordinates have a specific relationship between them. For example, if we required the number of blocks east must equal twice the number of blocks north, this describes some locations (6 blocks east, then 3 blocks north) but excludes others (3 blocks east, then 4 blocks north). A *curve* is the set of points that satisfy the given relationship. In this case, we'd require $x =$

$2y$, where we use the variable x to represent the horizontal coordinate and y the vertical coordinate.

Of particular interest are points whose coordinates are rational numbers (integers or fractions). For example, the point $(6, 3)$ is on the curve $x = 2y$, and since both 6 and 3 are rational numbers, then $(6, 3)$ is a *rational point* on the curve. On this curve, we can find other rational points easily. Since $x = 2y$, then as long as y is rational, x will be as well. Thus, we might pick a rational number for y, say, $y = \frac{11}{5}$ and compute the corresponding value of x: here $x = \frac{22}{5}$. The point $\left(\frac{22}{5}, \frac{11}{5}\right)$ is then a rational point on the curve $x = 2y$ (remember we're listing the x-coordinate first).

We can write down much more complicated equations involving x and y. An *elliptic curve* is an equation of the form $y^2 + pxy + qy = x^3 + rx^2 + ax + b$, where p, q, r, a, and b are real numbers . This describes so broad a class of curves that cryptographers focus on a smaller set, curves of the form $y^2 = x^3 + ax + b$, where a, b can be any integers except those that would make $4a^3 + 27b^2 = 0$.* Thus, $y^2 = x^3 - 3x + 2$ would not be an elliptic curve, since $4(-3)^3 + 27(2)^2 = 0$. On the other hand, $y^2 = x^3 - 3x + 4$ would be, since $4(-3)^3 + 27(4)^2 \neq 0$. A graph of the latter is shown, with the origin (O) and two points (A and B) labeled.

It is much harder to find a rational point on an elliptic curve. We might try our tactic of choosing a value of x or y and then using our formula to find the other coordinate. Thus, since $y^2 = x^3 - 3x + 4$, we can choose a value of x, and find the value of y^2; we could then find y. For example, if we tried $x = 0$, we'd find $y^2 = 0^3 - 3(0) + 4 = 4$, so y satisfies $y^2 = 4$. There are two numbers that solve this equation: $y = 2$ and $y = -2$. This gives us two rational points on the elliptic curve: $(0, 2)$ and $(0, -2)$.

We could try to find other rational points this way, but we'd quickly discover that *most* rational values of x don't produce rational points. For example, if we tried $x = 1$, we'd need $y^2 = (1)^3 - 3(1) + 4 = 2$. But however hard we try, we will never find a rational number y where $y^2 = 2$. We can try other values of x, and after a time, we *might* discover one: thus if we thought to try $x = \frac{9}{16}$, our y-value would have to satisfy $y^2 = \left(\frac{9}{16}\right)^3 - 3\left(\frac{9}{16}\right) + 4 = \frac{10,201}{4,096}$. Here we find that $y = \frac{101}{64}$ is a rational number that solves this equation (so is $y = -\frac{101}{64}$), and so we have two more rational points: $\left(\frac{9}{16}, \frac{101}{64}\right)$ and $\left(\frac{9}{16}, -\frac{101}{64}\right)$.

---

*The reason is that if $4a^3 + 27b^2 = 0$, the curve will intersect itself, which is a mathematically troublesome feature.

While finding *one* rational point is difficult, it turns out that if you have two rational points, finding more is very easy. If A and B are rational points on the curve, the line through them will intersect the elliptic curve at another rational point. It's convenient to focus on the *reflection* of this point (the point on the elliptic curve on the "opposite side" of the horizontal axis). Since this point was determined from points A and B, mathematicians designate this reflected point A + B (see Figure 12.1, A + B).

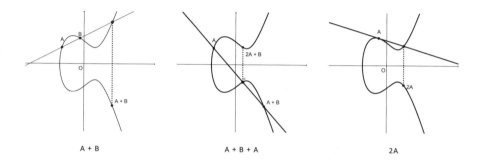

Figure 12.1. Elliptic Curve

But now A + B is a rational point, so the line between it and another rational point will intersect the curve at a third point. For example, we might draw the line between A and A + B; the reflection of this point will be A + A + B, which we designate 2A + B for short (see Figure 12.1, A + B + A). In a similar way, we can find 3A + B, 4A + B, and so on.

There are two special cases we need to consider. First, what about A + A? Addition relies on having a line between the two points, but in this case, the two points are the same. To resolve this, we'll use the *tangent line* through point A, which we can find using calculus; the tangent line intersects the curve at a third point, and the reflection of this point will be A + A, which we designate 2A (see Figure 12.1, 2A). Note that this means that once we find *one* rational point A on an elliptic curve, we can find another (namely, A + A) and then as many as we want.

The second special case occurs when two points lie on the same vertical line. Such a line will not intersect the curve at a third point, so in some sense the sum will be undefined. However, that's mathematically inconvenient, so we can resolve this by saying the two lines meet at the *point at infinity.* It's convenient to think of this as *the* point, which is at the top (and bottom) of *all* vertical lines.*

The point at infinity is identified as 0 for the following reasons. Consider any point A on the curve, and draw the vertical line through A. This vertical line passes through the point at infinity (since all vertical lines pass through the point at infinity), so we can use it to find A + 0. By the definition of how we add points on an elliptic curve, we'll find A + 0 by finding the third intersection point of the line and curve, then reflecting this point across the horizontal axis. A few pictures should convince you that A + 0 = A, so adding the point at infinity is like adding 0; it changes nothing.

Note that the above process of finding the point A + B from the points A, B is a purely geometric process. Given the elliptic curve, we draw the line through A and B and find a third point of intersection with the curve; we then reflect the point across the horizontal axis to find the point we designate A + B. We can take this *geometric* process and convert it into an *algebraic* formula: given the coordinates of the points A, B, we can find the coordinates of the point A + B. We claimed that the point A + B would be rational; the algebraic formulas that give the coordinates of the point A + B are what guarantee this claim is true.

In 1985, Victor Miller of IBM suggested that elliptic curves could be used as the basis of an encryption system; Neal Koblitz, at the University of Washington, independently made the suggestion in 1987 and resolved some of the practical difficulties (notably, how to find a point on the curve corresponding to the message being encrypted). However, neither connected the algorithm to a device, so the first to receive a patent for elliptic curve cryptography would be Ueli Maurer, professor of cryptography at the Swiss Federal Institute of Technology in Zurich. On September 8, 1992, Omnisec (a Swiss security corporation), received US Patent 5,146,500 for Maurer's system.

The transformation from an elliptic curve to a cryptographic system is a good example of mathematics in action. Suppose Alice wants to send a plaintext message m to Bob using ElGamal. To do so, they produce a shared key k and ciphertext km. Because they're working mod

---

*The point at infinity emerges naturally from projective geometry: two parallel railway tracks seem to meet at a point, so this point is the projection, onto the canvas, of the point at infinity.

N, the plaintext and key are numbers, and the ciphertext km is also a number.

Since the objects on an elliptic curve are points, we might begin moving ElGamal to elliptic curves by taking our message as a point M. Our key could be another point K, and our ciphertext the sum of the two points C = M + K. To decrypt, we find the additive inverse of K, which we designate –K, and find C + (–K). This produces elliptic curve ElGamal.

How do we produce a key? In ElGamal, the key was found by having Alice and Bob raise the public exponent to some secret power. We can do something similar on an elliptic curve. We might take some point A as our public base. If Alice chooses secret number m, she computes mA, which we define as the sum of m copies of A. For example, if Alice chooses m = 3, she computes

$$3A = A + A + A$$

and sends this point to Bob. Similarly, Bob chooses a secret number, say, n = 5, and computes

$$5A = A + A + A + A + A$$

Alice receives Bob's point 5A and computes 3(5A):

$$3(5A) = 5A + 5A + 5A$$

which, since 5A is itself the sum of 5 copies of A, will be the same as 15A: the sum of 15 copies of A.

Likewise, Bob receives Alice's 3A, and computes 5(3A)

$$5(3A) = 3A + 3A + 3A + 3A + 3A$$

Again, since 3A is the sum of 3 copies of A, Bob will have found 15A. As in ElGamal, Alice and Bob have now produced 15A at their respective computers without ever passing this value between them.

How about the additive inverse of 15A? This is even easier: the property of the additive inverse is that when we add 15A + (–15A), we get 0. Remember that we add two points on an elliptic curve by drawing a line between them; the reflection of the third point of intersection across the horizontal axis corresponds to the sum; and the point 0 corresponds to the point at infinity. Put together, this means that the additive inverse of 15A will just be the reflection of 15A across the horizontal axis, since if we draw the line between these two points, it will be vertical and pass through the point at infinity, whose reflection across the horizontal axis will be the point at infinity.

While this could be made to work, there's a problem. If we try this on an actual elliptic curve, we'll find that we end up with unwieldy fractional expressions. We can try to keep them as fractions, but the number of digits in the numerator and the denominator increases rapidly. Thus, if we take the point $(0, 2)$ as our point A on the elliptic curve $y^2 = x^3 - 3x + 4$, we'd find that A + A would be $\left(\frac{9}{16}, \frac{101}{64}\right)$. If we then added $\left(\frac{9}{16}, \frac{101}{64}\right)$ to itself (forming A + A + A + A, or 4A), we'd get the point $(-458{,}847/652{,}864,\ 1{,}266{,}176{,}783/527{,}514{,}112)$.

We can avoid these problems if we reduce our coordinates mod N. Since the formulas for finding A + B are defined in terms of addition, subtraction, multiplication and division, they easily translate into formulas mod N. For example, suppose we worked mod 1,187 (there are technical requirements for the modulus, which aren't important for our discussion). Then, if A is the point $(0, 2)$ on the elliptic curve $y^2 \equiv x^3 - 3x + 4 \pmod{1{,}187}$, then A + A would be the point $(965, 425)$, and the point 4A would be $(546, 310)$; more generally, since we're reducing the coordinates mod 1,187, then all of our coordinate values will be less than 1,187.

Thus, we might describe *elliptic curve ElGamal* as follows:

1. Alice and Bob agree on a suitable elliptic curve and modulus: for example, $y^2 \equiv x^3 - 3x + 4 \pmod{1{,}187}$. This is assumed to be public information.
2. They pick an integer solution to the elliptic curve: for example, $x = 446, y = 3$. This corresponds to some point A on the elliptic curve and is considered public information.
3. Alice picks a random number m and evaluates $mA = A + A + \ldots + A$ (the sum of m As), and sends this to Bob. As with Diffie-Hellman, this is Alice's half-key, and that mA will be some point on the curve.
4. Bob picks a random number n, evaluates nA, and sends this to Alice. Again, this is some point on the curve and corresponds to Bob's half-key.
5. Alice takes Bob's point nA and evaluates $m(nA) = nA + nA + \ldots + nA$, the sum of m points nA. This will be the point mnA, which will be the key.
6. Bob takes Alice's point mA and evaluates $n(mA)$. Again, he finds the key nmA.

7. To encrypt a message, Alice identifies it with a point M, then evaluates M + nmA, sending it to Bob.

8. Bob then computes M + nmA - nmA and recovers the original message.

As before, Eve's problem is finding m and n from Alice and Bob's half-keys mA and nA. By analogy with the DLP, this problem is known as the *elliptic curve discrete logarithm problem* (ECDLP) and is believed to be very hard to solve. More importantly, the only real similarity between the DLP and ECDLP is their names, so even if some method is found to solve the DLP (which would compromise ElGamal cryptosystems), elliptic curve cryptosystems would remain secure.

Elliptic Curves in Projective Coordinates

Both ElGamal and RSA require, at some point, raising a number to a very high power. In contrast, ECC only requires multiplication and addition. Thus, it's possible for an ECC cryptosystem to be significantly faster than a RSA or ElGamal system.

There's one catch. At some point in the addition of two points on an elliptic curve, it's necessary to find the multiplicative inverse of a number mod N. This can be done easily, using the Euclidean algorithm. However, it requires almost as many steps as the fast powering algorithm. This means that, in practice, an ECC cryptosystem is no faster than an ElGamal or RSA cryptosystem.

At the same time, any method of reducing the computation required to implement an ECC cryptosystem will make it more attractive. Certicom, a Canadian subsidiary of BlackBerry, has several particularly valuable patents in this regard and is generally regarded as being the primary holder of patents on elliptic curve cryptography. The first, US Patent 6,782,100, was issued on August 24, 2004. The patent describes a way to use *projective coordinates* to greatly reduce the computational requirements of an ECC cryptosystem.

Projective coordinates, also known as *homogeneous coordinates*, were invented in 1827 by the German mathematician August Ferdinand Möbius (1790–1868). In a projective coordinate system, the location of a point is specified by three coordinates (X, Y, Z). If Z is not equal to 0, then the "familiar" coordinates of the point can be recovered using $x =$

$\frac{X}{Z}$, $y = \frac{Y}{Z}$. Thus the point whose projective coordinates are $(12, 6, 2)$ corresponds to the point $(6, 3)$, since $x = \frac{12}{2} = 6$, $y = \frac{6}{2} = 3$.

The third coordinate might seem to be an unnecessary complication. However, projective coordinates have a number of useful features. For example, the reason they are also known as homogeneous coordinates is that they can transform a heterogeneous equation, one whose terms have different degree, into a homogeneous one.* Thus, in the equation $y = x^2$, the y term has degree 1 and the x term has degree 2. If we use projective coordinates, then $y = \frac{Y}{Z}$, $x = \frac{X}{Z}$, and our equation becomes $\frac{Y}{Z} = \frac{X^2}{Z^2}$, which we can rearrange to be $YZ^2 = X^2Z$. Here, the YZ^2 term has degree $1 + 2 = 3$, as does the X^2Z term.

Projective coordinates can speed computation as follows. Suppose we want to add two fractions, say, $\frac{1}{3}$ and $\frac{1}{6}$, and express the sum as a decimal. We can add them in two ways:

1. We convert $\frac{1}{3}$ into a decimal. Since $\frac{1}{3}$ gives us a nonterminating decimal, we have to decide how many places we're going to use. Suppose we decide to use eight decimal places. Then, $\frac{1}{3} = 0.33333333$.
2. We convert $\frac{1}{6}$ into a decimal. Again, we use eight decimal places: $\frac{1}{6} = 0.16666667$.
3. We add the decimal expressions: $0.33333333 + 0.16666667 = 0.50000000$.

If we do the computation this way, we have to perform two divisions, both of which must be carried out to eight decimal places.

As an alternative, we might do the following. We can view a number as a single ordinate (for example, the house number of a street), which can be converted into a projective coordinate by including a second value. Thus, instead of the number m, we have $(M, Z1)$, where $m = \frac{M}{Z1}$; likewise, instead of n, we have $(N, Z2)$, where $n = \frac{N}{Z2}$. To add $m + n$, we go through the following steps:

1. We find $P = M \times Z2 + N \times Z1$.
2. We find $Z3 = Z1 \times Z2$.
3. Then $m + n$ corresponds to $(P, Z3)$.

Since $(P, Z3)$ corresponds to the number $\frac{1}{16}$, we have found $m + n$

*The degree of a term is the sum of the exponents of the variables. Thus, the term x^3y^2 has degree $3 + 2 = 5$.

using multiplications, additions, and a *single* division. In this case, our fractions $\frac{1}{3}$ and $\frac{1}{6}$ become $(1, 3)$ and $(1, 6)$, and we have:

1. $P = 1 \times 6 + 1 \times 3 = 9$.
2. $Z3 = 3 \times 6 = 18$.
3. $m + n = (9, 18)$.

Our final step is computing $9 \div 18 = 0.5$.

This might seem a very roundabout way to calculate $\frac{1}{3} + \frac{1}{6}$. However, note that we've gone from having to perform three divisions to having to perform one. Since multiplication and addition can be performed far more rapidly than division, replacing even a single division with several multiplications and additions greatly reduces computation time.

Quantum Computers . . . and Beyond

All modern methods of cryptography rely on the belief that some problems are essentially impossible to solve without some additional information that can be kept secret from unauthorized users. RSA can be broken if we can factor a large number, and ElGamal can be broken if we can solve the discrete logarithm problem. However, the basis for this belief is tenuous: these problems are considered hard because no one has yet found an easy way to solve them.

A more serious threat centers around so-called *quantum computation*. We won't go into the details of quantum computing; it is sufficient to note that, for our purposes, a quantum computer will return something that is *probably* a solution. For example, consider the problem "Factor 56,153." A quantum computer might return the answer "241 × 233," which means that it is *probably* true that 56,153 = 241 × 233. What makes the "probably true" answer useful is that we can verify it. Similarly, if a quantum computer reports that the solution to $x^7 \equiv 146$ (mod 187) is probably $x = 5$, we can check for ourselves and see that $5^7 \equiv 146$ (mod 187).

There are two current difficulties with quantum computers. First, programming them to solve certain problems requires a solid understanding of quantum mechanics and computer science, so competent programmers are scarce. Nonetheless, algorithms already exist for factoring, solving the DLP, and even the ECDLP, so if a working quantum computer of sufficient power could be created, all of these cryptosystems would vulnerable.

The security of these cryptosystems is safe — for now — because the art of building quantum computers is in its infancy. Thus, even though we know how to use a quantum computer to factor a number, the largest number factored to date (as of mid-2017) is 291,311. Still, in anticipation that quantum computers of sufficient power will eventually be invented, cryptographers are investigating methods of *post-quantum cryptography*.

One promising direction is known as *lattice cryptography*. On May 21, 1996, computer scientist Miklós Ajtai (then at IBM) submitted a patent application on the use of lattices for solving the authentication problem. US Patent 5,737,425 would be granted to IBM on April 7, 1998.

Consider our coordinate system, but this time, imagine that we're only allowed to make certain moves. For example, we might limit ourselves to two types of moves. First, we might be able to move 127 blocks east *then* 8 blocks south: we'll call this S1. Second, we might be able to move 7 blocks east *then* 150 blocks north: we'll call this S2. We can also take these backward. Thus, the move −S1 would be reversing the moves of S1, taking us 127 blocks *west* and then 8 blocks *south*.

It's helpful to view these allowable moves as vectors. Remember that, to produce a vector, we need to decide on an order of the components. One possibility is to have a four-component vector, with the components giving the distances north, east, south, and west. But if we do this, we run into a problem. Consider two vectors such as [5, 0, 8, 0] and [0, 0, 3, 0]. The first vector describes going 5 blocks north and 8 blocks south. The second vector describes going 3 blocks south. It should be clear that both vectors describe the same move (namely, going 3 blocks south). But because they have different representations, they appear to be different.

Having different representations for the same quantity is inconvenient at best. It takes a moment's thought to realize that ¼ and 25% express the same fraction. To avoid this problem, we'll use two component vectors, with the first component describing the movement east-west, and the second describing the movement north–south. To distinguish between movement east and west and between movement north and south, we'll assign positive numbers to movement eastward or northward and negative to movement westward or southward.

Thus S1 (127 blocks east and then 8 blocks south) becomes the vector [127, −8], and S2 (7 blocks east and then 150 blocks north) becomes the vector [7, 150]. This notation has a further advantage. A quantity such as −S1 (reverse the directions of S1) can be expressed as

$$-1\,[127, -8] = [-127, 8]$$

where we use the scalar multiplication of vectors.

Just as we could use linear combinations of keyword vectors to describe a document, we can use linear combinations of direction vectors to specify a location. For example, consider the linear combination 2 S1 + S2; that's taking move S1 twice and then move S2. That is,

1. Making move S1 twice puts us 254 blocks east and 16 blocks south of our starting point.
2. Making move S2 puts us 7 blocks east (putting us 254 + 7 = 261 blocks east of our starting point) and 150 blocks north (which, since we were 16 blocks *south* of our starting point, would put us 134 blocks *north* of it).

Equivalently,

$$2\,[127, -8] + 1\,[7, 150] = [261, 134]$$

where we use both scalar multiplication and vector addition.

Because we're constrained in the moves we can make, we'll find there are some places we can get to and some places we can't. As the preceding shows, we can form the vector [261, 134], which takes us to the point (261, 134). However, we can't get to the point (1, 1), since no linear combination will form the vector [1, 1]. The locations we can get to form a *lattice*, and the moves we're allowed to make, such as S1 and S2, are the *basis vectors* of the lattice.

Lattices lead to two problems. The *shortest vector problem* (SVP) is given the basis vectors for a lattice, find the lattice point closest to the origin; this corresponds to finding the shortest vector we can produce as a linear combination. The *closest vector problem* (CVP) is similar. Given the basis vectors for a lattice and some point, find the lattice point closest to the given point. Both are believed to be hard problems. Moreover, unlike factoring, we can't easily verify a proposed solution, which limits the usefulness of quantum computers in solving this problem.

Actually, if you look at the given vectors, it's easy to solve the SVP, since the two vectors are nearly perpendicular. In particular, S1 *mostly* takes you east, while S2 *mostly* takes you north. Common sense suggests that the SVP could be solved by *either* moving S1, moving S2, or moving S1 + S2. More generally, if all the basis vectors are nearly perpendicular (so that our basis vectors are *quasi-orthogonal*), SVP is easy to solve.

Now consider another set of basis vectors: P1 = [204, 1642] and P2 = [415, 3434]. What might not be apparent is that every lattice point we can get to using S1 and S2 can also be arrived at using P1 and P2. However, because P1 and P2 are very nearly parallel (they both move us some distance east and about eight times as far north), it's much harder to solve the SVP: P1 and P2 form a "bad" basis for the lattice.

Ajtai's patent describes how we can produce such a bad basis from a good basis. This in turn can be used to solve the authentication problem as follows. As with Diffie-Hellman, Bob publishes the bad basis P1, P2 for the lattice. Since solving the SVP is hard for those who don't know the good basis S1, S2, then Bob can prove his identity by finding the lattice point closest to the origin.

If Alice challenges Bob to prove his identity, he can solve the SVP and send the solution, which will be the point (127, –8), to Alice.

Because Alice doesn't know the good basis, she can't actually verify that (127, –8) solves the SVP. But at the very least, she can verify that it's a lattice point. In particular, there must be a way to get there using the directions P1 and P2, so there must be integer values a, b that produce the vector [127, –8]. This gives the equation:

$$a[204, 1642] + b[415, 3434] = [127, -8]$$

This corresponds to solving the system of equations

$$204a + 415b = 127$$
$$1642a + 3434b = -8$$

This is a very easy equation to solve, and Alice finds that $a = 23$, $b = -11$.

There are two problems with this approach. First, Alice doesn't know the solution to the SVP, so she has to convince herself that Bob's solution is the correct one. A bigger problem is that *once Bob has solved the SVP for this lattice, he must change the lattice.* Even though the proof revealed nothing about the good basis for the lattice, the original lattice is useless for future authentications.

Keeping Secrets in a Boiler Factory

The problem of using lattices as the basis for authentication and cryptography was actually solved between the time Ajtai applied for his patent and the time he received it. In 1997, computer scientists Oded

Goldreich, of the Weizmann Institute of Science in Israel; and Shafi Goldwasser and Shai Halevi, of MIT, published a paper outlining a cryptographic method now known as GGH.

Conceptually, GGH is based on the idea that a little noise can render a conversation unintelligible, unless you have some idea of what the conversation is about. For example, suppose Alice and Bob are on a noisy factory floor. Bob asks Alice when a task will be completed. Her response sounds like "Today, I'm thirsty." But since Bob knows Alice is responding to a question about *when*, he interprets Alice's statement as "Two days from Thursday." In contrast Eve, who doesn't know what they're talking about, will remain baffled by Alice's answer. In the same way, the necessity of using lattice points limits the conversational topics.

GGH works as follows:

1. Alice chooses a private, quasi-orthogonal basis for a lattice.
2. She then constructs a "bad" public basis for the lattice.
3. Bob encrypts a sequence of numbers by making them coefficients for a linear combination of the basis vectors; this determines some point Q on the lattice.
4. Bob then chooses a random "noise" vector N and finds a nearby point Q' = Q + N.
5. Bob sends the nearby point Q' to Alice.
6. Alice solves the CVP and finds Q.
7. Alice then uses the public basis to recover Bob's number.

As with RSA, this method can be used for encryption or for authentication.

Suppose Bob wants to verify Alice's identity. First, he picks some random lattice point Q, based on the "bad" public basis P1, P2. For example, he might choose 5 P1 + 11 P2 = [5585, 45894]. He then adds some "noise" and changes the coordinates slightly: for example, he might change them to the lattice point (5590, 45980).

Now he challenges Alice to find the closest lattice point to (5584, 45981). Alice, who knows the quasi-orthogonal basis, tries to find a linear combination of S1 and S2 that will take her as close as possible to (5590, 45980). Thus she must solve

$$a\ S1 + b\ S2 = [5590, 45980]$$

She finds $a \approx 27.04$ and $b \approx 307.98$. Since a, b have to be integers, she rounds them to $a = 27$ and $b = 308$ and finds 27 S1 + 308 S2 = (5585, 45984), which is Bob's original point. The fact that she found Bob's original point is proof that she knows the private basis — and thus proof that she is who she says she is.

What about Eve? Since she only knows the public basis, she tries to find a linear combination of P1 and P2 that puts her close to (5590, 45980). Thus, she tries to solve

$$a' \text{ P1} + b' \text{ P2} = [5590, 45980]$$

She finds $a' \approx 5.99$ and $b' \approx 10.53$. Again, since a', b' need to be integers, Eve rounds them to $a' = 6$, $b' = 11$, corresponding to the vector 6 P1 + 11 P2 = [5789, 47626], suggesting the closest point is (5789, 47626). But this is far away from Bob's original (5585, 45984). Thus, Bob concludes that Eve does not know the private basis S1, S2.

In contrast to Ajtai's patent, this approach can be reused. If Bob needs to verify Alice's identity in the future, he should choose a different lattice point. Moreover, as with the Diffie-Hellman authentication approach, Bob can't use Alice's answer to impersonate her, since it's unlikely that someone else would pick the same lattice point he did.

Moreover, the system can be easily converted to a cryptographic system. Suppose Bob wants to send a credit card number to Alice. He breaks the number into parts: for example, the number 0511 can be broken into 05 and 11. These parts can then be used as coefficients to specify a lattice point, and then noise can be added as above before sending the encrypted value to Alice. Thus, Bob computes 5 P1 + 11 P2 = [5585, 45894] then adds a little noise to produce the point (5590, 45980). He sends this point to Alice.

As before, Alice solves the CVP and finds (5585, 45984) is the closest lattice point. Since Bob obtained this using the public basis, she can then recover Bob's original numbers by solving

$$x \text{ S1} + y \text{ S2} = [5585, 45984]$$

This gives her $x = 5$, $y = 11$.

In contrast, Eve doesn't know the point Bob actually found, so she uses the noisy value and tries to solve $a' \text{ P1} + b' \text{ P2} = [5590, 45980]$ directly, finding $a' \approx 5.99$ and $b' \approx 10.53$, and concluding Bob was trying to send the numbers 6 and 11, and "recovering" the incorrect credit card number 0611.

GGH might have replaced RSA immediately but for two problems. First, if you have a quasi-orthogonal basis for the lattice, solving SVP or CVP is not too hard. There is a standard method for transforming any basis into a quasi-orthogonal basis, which is effective if the number of basis vectors isn't too large. Any secure implementation of GGH requires having a large number of basis vectors. Current studies suggest that a reasonable amount of security can be had if there are more than 400 basis vectors. But this leads to a second problem. Using GGH with that many basis vectors requires far more computations than a similarly secure RSA system. Thus, secure GGH implementations are impractical, while practical GGH implementations are insecure.

However, just as elliptic curves could be moved from geometry to a number system mod N, we can move the lattice problem into another type of number system known as a *polynomial ring*. In 1996, Brown University mathematicians Jeffrey Hoffstein, Jill Pipher, and Joseph Silverman developed a method of lattice cryptography based on polynomial rings, and on October 2, 2001, they received US Patent 6,298,137 for a cryptographic system known as NTRU (which allegedly stands for "Number Theorists R Us," among other claims). Together, with Daniel Lieman, they founded NTRU Cryptosystems, which would be acquired by Security Innovation in 2009.

NTRU defies easy explanation, though preliminary studies suggest it provides very strong security. Moreover, there has been very little progress on finding quantum algorithms for solving lattice problems, so they are likely to be secure even after the development of working quantum computers of sufficient complexity. Even better, a working quantum computer could greatly speed the implementation of lattice cryptographic schemes, making them practical for day-to-day use. We may yet lose the Red Queen's Race but not for lack of trying.

Epilogue

A patent for a claimed invention may not be obtained . . .
if the differences between the claimed invention and the prior art are
such that the claimed invention as a whole would have been obvious
before the effective filing date of the claimed invention to a person having
ordinary skill in the art to which the claimed invention pertains.

35 USC § 103

On June 19, 2014, the United States Supreme Court ruled in *Alice Corp. v. CLS Bank International* that simply tying an application to a computer was *not* enough for patent eligibility. Less than a month later, on July 11, 2014, the Third Circuit Court went further, in *Digitech Image Technologies v. Electronics for Imaging Inc.* The court ruled that "a process that employs mathematical algorithms to manipulate existing information to generate additional information is not patent eligible." As result, the Digitech patent would be reevaluated and voided by the patent office on February 14, 2017.

Because nearly every application of mathematics will require tying it to a computer, and because *all* mathematically based patents use algorithms to manipulate existing information to generate additional information, the *Digitech* ruling could potentially invalidate *every* patent based on mathematics.

Subsequently, various rulings have tried to identify when such a device is patentable. The most important of these was *Enfish v. Microsoft*, decided May 12, 2016, also by the Third Circuit, which established some conditions under which software (and, thus, computer implementation of mathematical algorithms) might be patentable.

While the *Enfish* ruling offers some clarity, the greatest risk in allowing a patent based on mathematics is that mathematics is so readily generalizable that a single patent could give its owners a stranglehold over *every* industry that uses mathematics, which is to say all of them. As we've seen, the method of comparing two documents can be used to compare two images, suggest career trajectories, or find potential lifemates. The algorithm for ranking web pages can be converted into a method of identifying crucial customers; a procedure for disassembling LEGO models can be used to plan computer network capacities; a function that evaluates password strength can be used to compress data or to plan an educational curriculum; and so on. To a mathematician, transforming from one field to another is second nature: all of these would be obvious to "those skilled in the art."

It is imperative, if patents are to encourage innovation, that these uses not be covered by a single, broad patent. Equally important, an inventor should not be able to preempt an entire industry with an airy reference to "other uses, obvious to those skilled in the art."

To that end, we make the following suggestions for US patent policy.

First, 35 USC § 112 requires a patentable device to work as claimed. We suggest that all patents based on mathematical algorithms and formulas be subject to strict scrutiny with regards to this requirement. In particular, the formula or algorithm should be clearly defined and evidence provided that the specified formula work as claimed. Not only will this serve to limit overly broad claims, but it will also make it easier to determine when a significant improvement has been made to an existing invention.

Second, under current guidelines, those seeking to become patent agents who work with inventors to expedite the application process and serve as liaisons between the patent office and the inventor, must submit evidence of scientific and technical expertise. Currently, mathematics coursework is expressly excluded as evidentiary. However, as more and more patents rely on mathematical algorithms, this exclusion becomes less and less reasonable. We suggest that appropriate-level coursework in mathematics be acceptable as evidence of scientific and technical expertise.

Finally, the position of patent examiner requires a degree in engineering or science but not in mathematics. We suggest that the patent office also consider candidates holding degrees in mathematics. Again, as patents based on mathematical algorithms are becoming more common, we argue that it is necessary that the patent office gain expertise and the ability to distinguish between what is truly an innovative and patent-worthy application of mathematics and what is a simple extension of a well-understood concept.

Aikins, Herbert Augustin. Educational Appliance, US Patent 1,050,327, filed February 27, 1911, and issued January 14, 1913.

Ajtai, Miklós. Cryptosystem Employing Worst-Case Difficult-to-Solve Lattice Problem, US Patent 5,737,425, filed May 21, 1996, and issued April 7, 1998.

Altman, George G. Apparatus for Teaching Arithmetic, US Patent 588,371, filed October 6, 1896, and issued August 17, 1897.

Alward, Herbert Lewis; Meehan, Timothy Erickson; Straub, James Joseph, III; Hust, Robert Michael; Hutchinson, Erik Watson; Schmidt, Michael Patrick. Continuous User Identification and Situation Analysis with Identification of Anonymous Users through Behaviormetrics, US Patent 8,843,754, filed September 17, 2007, and issued September 23, 2014.

Anderson, Gary E.; Ramsey, Mark S.; Selby, David A. Social Network Analysis for Churn Prediction, US Patent 8,843,431, filed January 16, 2012 and issued September 23, 2014.

Argaiz, Joseluis Iribarren. Method and System for Managing Customer Network Value, US Patent 7,941,339, filed November 17, 2005, and issued May 10, 2011.

Atreya, Avinash; Elkan, Charles. "Latent Semantic Indexing (LSI) Fails for TREC Collections." *ACM SIGKDD Explorations Newsletter*, Vol. 12, No. 2 (2010), 5–10.

Bamhart, Randall. Entropy-Based Sequences of Educational Modules, filed November 27, 2013, and issued November 4, 2014.

Benjamin, Ludy, Jr. "A History of Teaching Machines," *American Psychologist*, Vol. 43, No. 9 (1988), 703–712.

Berry, Michael; Drmac, Zlatko; Jessup, Elizabeth. "Matrices, Vector Spaces, and Information Retrieval." *SIAM Review*, Vol. 41, No. 2 (1999), 335–362.

Bharat, Krishna Asur; Henzinger, Noika R. Method for Ranking Documents in a Hyperlinked Environment Using Connectivity and Selective Content Analysis, US Patent 6,112,203, filed April 9, 1998, and issued August 29, 2000.

Blose, Andrew; Stubler, Peter; Manico, Joseph. Automatic Appeal Measurement System, US Patent 8,369,582, filed June 24, 2010, and issued February 5, 2013.

Bojinov, Hristo; Sanchez, Daniel; Reber, Paul; Boneh, Dan; Lincoln, Patrick. "Neuroscience Meets Cryptography: Crypto Primitives Secure against Rubber Hose Attacks." *Communications of the ACM*, Vol. 57, No. 5 (2014), 110–118.

Borniger, Jeremy; Chaudhry, Adeel; Muehlenbein, Michael. "Relationships among Musical Aptitude, Digit Ratio, and Testosterone in Men and Women," *PLOS ONE*, Vol. 8, No. 3 (2013), e57637.

Brin, S., Page, L. "The Anatomy of a Large-Scale Hypertextual Web Search Engine" (paper presented at Seventh International World-Wide Web Conference [WWW 1998], Brisbane, Australia, April 14–18, 1998).

Brown, Jacqueline Johnson; Reingen, Peter H. "Social Ties and Word-of-Mouth Referral Behavior." *Journal of Consumer Research*, Vol. 14 (1987), 350–362.

Buckwalter, J. Galen; Carter, Steven R.; Forgatch, Gregory T.; Parsons, Thomas D.; Warren, Neil Clark. Method and System for Identifying People Who Are Likely to Have a Successful Relationship, US Patent 6,735,568, filed August 10, 2000, and issued May 11, 2004.

Burr, William; Dodson, Donna; Newton, Elaine; Perlner, Ray; Polk, W. Timothy; Gupta, Sarbari; Nabbus, Emad. *Electronic Authentication Guideline.* National Institute of Standards and Technology, 2013.

Bush, Vannevar. "As We May Think." *The Atlantic,* July 1945.

Buss, David; Shackelford, Todd; Kirkpatrick, Lee; Larsen, Randy. "A Half Century of Mate Preferences: The Cultural Evolution of Values." *Journal of Marriage and Family,* Vol. 63 (2001), 491–503.

Byrne, Donn; Ervin, Charles; Lamberth, John, "Continuity between the Experimental Study of Attraction and Real-Life Computer Dating." *Journal of Personality and Social Psychology,* Vol. 16, No. 1 (1970), 157–165.

Canright, Geoffrey; Engø-Monsen, Kenth. "Roles in Networks." *Science of Computer Programming,* Vol. 53 (2004), 195–214.

Canright, Geoffrey; Engø-Monsen, Kenth; Weltzien, Asmund. Method for Managing Networks by Analyzing Connectivity, US Patent 7,610,367, filed July 14, 2005, and issued October 27, 2009.

Cantor, Allison; Eslick, Andrea; Marsh, Elizabeth; Bjork, Robert; Bjork, Elizabeth. "Multiple-Choice Tests Stabilize Access to Marginal Knowledge." *Memory and Cognition,* Vol. 43 (2015), 193–205.

Carnevale, Anthony; Smith, Nicole; Strohl, Jeff. *Projections of Jobs and Education Requirements through 2018.* Georgetown University Center on Education and the Workforce, June 2010.

Chakraborty, Dipanjan; Dasgupta, Koustuv; Mukherjea, Sougata; Nanavati, Amit; Singh, Rahul; Viswanthan, Balaji. System and Computer Program Product for Predicting Churners in a Telecommunication Network, US Patent 8,249,231, filed January 28, 2008, and issued August 21, 2012.

Chartier, Tim. *When Life Is Linear: From Computer Graphics to Bracketology.* MAA Press, 2015.

Chevreul, M. E., *De la Loi du Contraste Simultané des Couleurs.* Pitois-Levrault, 1839.

Coates, John M.; Gurnell, Mark; Rustichini, Aldo. "Second-to-Fourth Digit Ratio Predicts Success among High-Frequency Financial Traders." *Proceedings of the National Academy of Sciences,* Vol. 106, No. 2 (2009), 623–628.

Cohen, Nathan. Fractal Antenna Ground Counterpoise, Ground Planes, and Loading Elements, US Patent 6,140,975, filed November 7, 1997, and issued October 31, 2000.

Coleman, James; Katzu, Elihu; Menzel, Herbert. "The Diffusion of an Innovation among Physicians." *Sociometry,* Vol. 20, No. 4 (1957), 253–270.

Coleman, Mark; Barbato, Darrel; Huang, Lihu. System and Method for Determining an Insurance Premium Based on Complexity of a Vehicle Trip, US Patent 8,799,035, filed September 17, 2013, and issued August 5, 2014.

Collins, Thomas; Hopkins, Dale; Langford, Susan; Sabin, Michael. Multiprime RSA Public Key Cryptosystem, US Patent 7,231,040, filed October 26, 1998, and issued June 12, 2007.

Conover, Lloyd H. Tetracycline. US Patent 2,699,054, filed October 9, 1953 and issued January 11, 1955.

Cordara, Giovanni; Francini, Gianluca; Lepsoy, Skjalg; Porto Buarque de Gusmao, Pedro. Method and System for Comparing Images, US Patent 9,008,424, filed January 25, 2011, and issued April 14, 2015.

Dasgupta, Koustuvu; Singh, Rahul; Viswanthan, Balaji; Chakraborty, Dipanjan; Mukherjea, Sougata; Nanvati, Amit; Joshi, Anupam. "Social Ties and Their Relevance to Churn in Mobile Telecom Networks" (paper presented at 11th Conference on Extending Database Technology, Nantes, France, March 25–29, 2008).

Deerwester, Scott; Dumais, Susan; Furnas, George; Harshman, Richard; Landauer, Thomas; Lochbaum, Karen; Streeter, Lynn. Computer Information Retrieval Using Latent Semantic Structure, US Patent 4,839,853, filed September 15, 1988, and issued June 13, 1989.

Deerwester, Scott; Dumais, Susan; Harshman, Richard. "Indexing by Latent Semantic Analysis." *Journal of the American Society for Information Science*, Vol. 41, No. 6 (1990), 391–407.

Diffie, Whitfield; Hellman, Martin E. "New Directions in Cryptography." *IEEE Transactions on Information Theory*, Vol. IT-22, No. 6 (1976), 644–654.

Dorigo, Marco. "Ant Colony System: A Cooperative Learning Approach to the Traveling Salesman Problem." *IEEE Transactions on Evolutionary Computation*, Vol. 1, No. 1 (1997), 53–66.

Drennan, Ricky J.; Dubey, Sharmistha; Ginsberg, Amanda; Roberts, Anna M.; Smith, Marty L.; Woodbye, Stanley E., Jr. System and Method for Providing Enhanced Matching Based on Question Responses, US Patent 8,195,668, filed September 5, 2008, and issued June 5, 2012.

Durand, Pierre E.; Low, Michael D.; Stoller, Melissa K. System for Data Collection and Matching Compatible Profiles, US Patent 6,272,467, filed January 16, 1997, and issued August 7, 2001.

Durhuus, Bergfinnur; Eilers, Søren. "On the Entropy of LEGO." *Journal of Applied Mathematics and Computing*, Vol. 45, No. 1 (2014), 433–448.

Eastwick, Paul; Finkel, Eli. "Sex Differences in Mate Preferences Revisited: Do People Know What They Initially Desire in a Romantic Partner?" *Journal of Personality and Social Psychology*, Vol. 94, No. 2 (2008), 245–264.

Elgamal, Taher. "A Public Key Cryptosystem and a Signature Scheme Based on Discrete Logarithms." *IEEE Transactions on Information Theory*, Vol. IT-31, No. 4 (1985), 469–472.

Enss, Chris. "Getting Personal on the Frontier." *Wild West*, February 2015, 44–51.

Finkel, Eli J.; Eastwick, Paul; Karney, Benjamin; Reis, Harry; Sprecher, Susan. "Online Dating: A Critical Analysis from the Perspective of Psychological Science." *Psychological Science in the Public Interest*, Vol. 13, No. 1 (2012), 3–66.

Fiore, Andrew; Donath, Judith. "Homophily in Online Dating: When Do You Like Someone Like Yourself?" (paper presented at CHI Conference on Human Factors in Computing Systems, Portland, Oregon, April 2–7, 2005).

Gaines, R. Stockton; Lisowski, William; Press, S. James; Shapiro, Norman. *Authentication by Keystroke Timing: Some Preliminary Results*. Rand Corporation, 1980.

Galen, William; Warnakulasooriya, Rasil. Fractal-Based Decision Engine for Intervention, US Patent 8,755,737, filed December 24, 2012, and issued June 17, 2014.

Goldreich, Oded; Goldwasser, Shafi; Halevi, Shai. "Public-Key Cryptosystems from Lattice Reduction Problems." In *Advances in Cryptology*, ed. B. S. Klaiski Jr. Springer-Verlag, 1997.

Gower, Rebecca; Heydtmann, Agnes; Petersen, Henrik. "LEGO: Automated Model Construction." In *32nd European Study Group with Industry, Final Report*. Department of Mathematics, DTU, 1998, 81–94.

Granovetter, Mark. "The Strength of Weak Ties." *American Journal of Sociology*, Vol. 78, No. 6 (1973), 1360–1380.

Gunes, Hatice; Piccardi, Massimo. "Assessing Facial Beauty through Proportional Analysis by Image Processing and Supervised Learning." *International Journal of Human-Computer Studies*, Vol. 64 (2006), 1184–1199.

Hardy, G. H. *A Mathematician's Apology*. Cambridge University Press, 1940.

Hellman, Martin E.; Diffie, Bailey W.; Merkle, Ralph C. Cryptographic Apparatus and Method, US Patent 4,200,770, filed September 6, 1977, and issued April 29, 1980.

Hill, Reuben. "Campus Values in Mate Selection," *Journal of Home Economics*, 37 (1945), 554–558.

Hisasue, Shin-ichi; Sasaki, Shoko; Tsukamoto, Taiji; Horie, Shigeo. "The Relationship between Second-to-Fourth Digit Ratio and Female Gender Identity." *Journal of Sexual Medicine*, Vol. 9 (2012), 2903–2910.

Hoffstein, Jeffrey; Pipher, Jill; Silverman, Joseph H. Ring-Based Public Key Cryptosystem Method, US Patent 6,298,137, filed April 5, 2000, and issued October 2, 2001.

House, Lisa; House, Mark; Mullady, Joy. "Do Recommendations Matter? Social Networks, Trust, and Product Adoption." *Agribusiness*, Vol. 24, No. 3 (2008), 332–341.

Jakobsen, Jakob Sprogoe; Ernstvang, Jesper; Kristensen, Ole; Allerelli, Jacob. Automatic Generation of Building Instructions for Building Element Models, US Patent 8,374,829, filed September 15, 2009, and issued February 12, 2013.

Judy, Richard W. Method for Statistical Comparison of Occupations by Skill Sets and Other Relevant Attributes, US Patent 8,473,320, filed December 15, 2011, and issued June 25, 2013.

Kamvar, Sepandar; Haveliweala, Taher; Jeh, Glen. Method for Detecting Link Spam in Hyperlinked Databases, US Patent 7,509,344, filed August 18, 2004, and issued March 24, 2009.

Kawale, Jaya; Pal, Aditya; Srivastava, Jaideep. "Churn Prediction in MMORPGs: A Social Influence Based Approach" (paper presented at the International Conference on Computational Science and Engineering, Vancouver, Canada, August 29–31, 2009).

Keener, James P. "The Perron-Frobenius Theorems and the Ranking of Football Teams." *SIAM Review*, Vol. 35, No. 1 (1993), 80–93.

Kim, Hyoungchol. Process for Designing Rugged Pattern on Golf Ball Surface, US Patent 8,301,418, filed September 3, 2009, and issued October 30, 2012.

Kirsch, Steven T. Document Retrieval over Networks Wherein Ranking and Relevance Scores Are Computed at the Client for Multiple Database Documents, US Patent 5,659,732, filed May 17, 1995, and issued August 19, 1997.

Kiss, Christine; Bichler, Martin. "Identification of Influencers—Measuring Influence in Customer Networks." *Decision Support Systems*, Vol. 46 (2008), 233–253.

Kleinberg, Jon Michael. Method and System for Identifying Authoritative Information Resources in an Environment with Content-Based Links between Information Resources, US Patent 6,112,202, filed March 7, 1997, and issued August 29, 2000.

Langville, Amy N.; Meyer, Carl D. *Google's PageRank Algorithm and Beyond*. Princeton University Press, 2012.

Langville, Amy; Meyer, Carl D. "A Survey of Eigenvector Methods for Web Information Retrieval." *SIAM Review*, Vol. 47, No. 1 (2005), 135–161.

Langville, Amy N.; Meyer, Carl D. *Who's #1? The Science of Rating and Ranking*. Princeton University Press, 2012.

Lee, Morris. Image Comparison Using Color Histograms, US Patent 8,897,553, filed December 13, 2011, and issued November 25, 2014.

Leinster, Murray. "A Logic Named Joe." In *The Best of Murray Leinster*. Del Rey, 1978.

Leung, Gilbert; Duan, Lei; Pavlovski, Dmitri; Chan, Su Han; Tsioutsiouliklis, Kostas. Detection of Undesirable Web Pages, US Patent 7,974,970, filed October 9, 2008, and issued July 5, 2011.

Lineaweaver, Sean; Wakefield, Gregory. Using a Genetic Algorithm Employing Dynamic Mutation, US Patent 8,825,168, filed September 10, 2009, and issued September 2, 2014.

Linson, Jesse. Public Key Cryptographic Methods and Systems, US Patent, 7,760,872, filed December 30, 2004, and issued July 20, 2010.

Maurer, Ueli. Public Key Cryptographic System Using Elliptic Curves over Rings, US Patent 5,146,500, filed March 22, 1991, and issued September 8, 1992.

McEntire, Lauren; Dailey, Lesley; Osburn, Holly; Mumford, Michael. "Innovations in Job Analysis: Development and Application of Metrics to Analyze Job Data." *Human Resource Management Review*, Vol. 16 (2006), 310–323.

Meltzer, Andrea L.; McNulty, James K.; Jackson, Grace; Karney, Benjamin R. "Sex Differences in the Implications of Partner Physical Attractiveness for the Trajectory of Marital Satisfaction." *Journal of Personality and Social Psychology*, Vol. 106, No. 3 (2014), 418–428.

Morris, Robert; Thompson, Ken. "Password Security: A Case History." *Communications of the ACM*, Vol. 22, No. 11 (1979), 594–597.

Mullany, Michael. System and Method for Determining Relative Strength and Crackability of a User's Security Password in Real Time, US Patent 7,685,431, filed March 20, 2000, and issued March 23, 2010.

Page, Lawrence. Method for Node Ranking in a Linked Database, US Patent 6,285,999, filed January 9, 1998, and issued September 4, 2001.

Petrina, Stephen. "Sidney Pressey and the Automation of Education, 1924–1934." *Technology and Culture*, Vol. 45, No. 2 (2004), 305–330.

Phadke, Chitra; Uzunalioglu, Huseyin; Menditratta, Veena; Kushnir, Dan; Doran, Derek. "Prediction of Subscriber Churn Using Social Network Analysis," *Bell Labs Technical Journal*, Vol. 17, No. 4 (2013), 63–76.

Phadke, Chitra; Uzunalioglu, Huseyin; Menditratta, Veena; Kushnir, Dan; Doran, Derek. System and Method for Generating Subscriber Churn Predictions, US Patent 8,804,929, filed October 30, 2012, and issued August 12, 2014.

Prager, John Martin. System and Method for Categorizing Objects in Combined Categories, US Patent 5,943,670, filed November 21, 1997, and issued August 24, 1999.

Pressey, Sidney. Machine for Intelligence Tests, US Patent 1,670,480, filed January 30, 1926, and issued May 22, 1928.

Rabinovitch, Peter; McBride, Brian. Simulated Annealing for Traffic Matrix Estimation, US Patent 7,554,970, filed October 13, 2004, and issued June 30, 2009.

Rivest, Ronald; Shamir, Adi; Adleman, Leonard. Cryptogrpahic Communications System and Method, US Patent 4,405,829, filed December 14, 1977, and issued September 20, 1983.

Salmon, D. *Elements of Computer Security.* Springer-Verlag, 2010.

Salton, G.; Wong, A.; Yang, C. S. "A Vector Space Model for Automatic Indexing." *Communications of the ACM*, Vol. 18, No. 11 (1975), 613–620.

Salton, Gerard. *A Theory of Indexing.* Cornell University, 1974.

Salton, Gerard; Buckley, Christopher. "Term-Weighting Approaches in Automatic Text Retrieval." *Information Processing and Management*, Vol. 24, No. 5 (1988), 513–523.

Schuetze, Hinrich. Document Information Retrieval Using Global Word Co-occurrence Patterns, US Patent 5,675,819, filed June 16, 1994, and issued October 7, 1997.

Shannon, Claude. "A Mathematical Theory of Communication." *The Bell System Technical Journal*, Vol. 27 (July 1948), 379–423.

Sharma, Ravi K. Histogram Methods and Systems for Object Recognition, US Patent 8,767,084, filed August 19, 2011, and issued July 1, 2014.

Shay, Richard; Komanduri, Saranga; Kelley, Patrick; Leon, Pedro; Mazurek, Michelle; Bauer, Lujo; Christin, Nicolas; Cranor, Lorrie. "Encountering Stronger Password Requirements: User Attitudes and Behaviors" (paper presented at Symposium on Usable Privacy and Security, Redmond, Washington, July 14–16, 2010).

Shen, Zeqian; Sundaresan, Neelakantan. Method and System for Social Network Analysis, US Patent 8,473,422, filed November 30, 2010, and issued June 25, 2013.

Shuster, Brian. Method, Apparatus, and System for Directing Access to Content on a Computer Network, US Patent 6,389,458, filed October 30, 1998, and issued May 14, 2002.

Shute, Valerie; Hansen, Eric; Almond, Russell. Method and System for Designing Adaptive, Diagnostic Assessments, US Patent 7,828,552, filed February 22, 2006, and issued November 9, 2010.

Skinner, B. F. "Teaching Machines." *Science*, Vol. 128, No. 3330 (1958), 969–977.

Skinner, Burrhus Frederic. Teaching Machine, US Patent 2,846,779, filed May 12, 1955, and issued August 12, 1958.

Soulie-Fogelman, Françoise. Method and System for Selecting a Target with Respect to a Behavior in a Population of Communicating Entities, US Patent 8,712,952, filed November 15, 2011, and issued April 29, 2014.

Suzuki, J. "For Better Security, Try a Random-Point Password Sequence." *Communications of the ACM*, Vol. 57, No. 12 (2015), 9.

Svenonius, Elaine. "An Experiment in Index Term Frequency." *Journal of the American Society for Information Science*, Vol. 23, No. 2 (1972), 100–121.

Swain, Michael J.; Ballard, Dana H. "Color Indexing." *International Journal of Computer Vision*, Vol. 7, No. 1 (1991), 11–32.

Tsai, Sam; Chen, David; Takacs, Gabriel; Chadrasekhar, Vijay; Vedantham, Ramakrishna; Grzeszczuk, Radek; Girod, Bernd. "Fast Geometric Re-ranking for Image-Based Retrieval" (paper presented at IEEE International Conference, Piscataway, New Jersey, September 26–29, 2010).

Turney, Peter D.; Pantel, Patrick. "From Frequency to Meaning: Vector Space Models of Semantics." *Journal of Artificial Intelligence Research*, Vol. 37 (2010), 141–188.

Vanstone, Scott; Mullin, Ronald; Antipa, Adrian; Gallant, Robert. Accelerated Finite Field Operations on an Elliptic Curve, US Patent 6,782,100, filed October 2, 2000, and issued August 24, 2004.

Warnakulasooriya, Rasil; Galen, William. "Categorizing Students' Response Patterns Using the Concept of Fractal Dimension." In *Educational Data Mining*, 2012.

Wheeler, Dan Low. Pattern Entropy Password Strength Estimator, US Patent 8,973,116, filed December 19, 2012, and issued March 3, 2015.

Wong, Ken Kay. "Getting What You Paid For: Fighting Wireless Customer Churn with Rate Plan Optimization." *Database Marketing and Customer Strategy Management*, Vol. 18, No. 2 (2011), 73–82.

INDEX